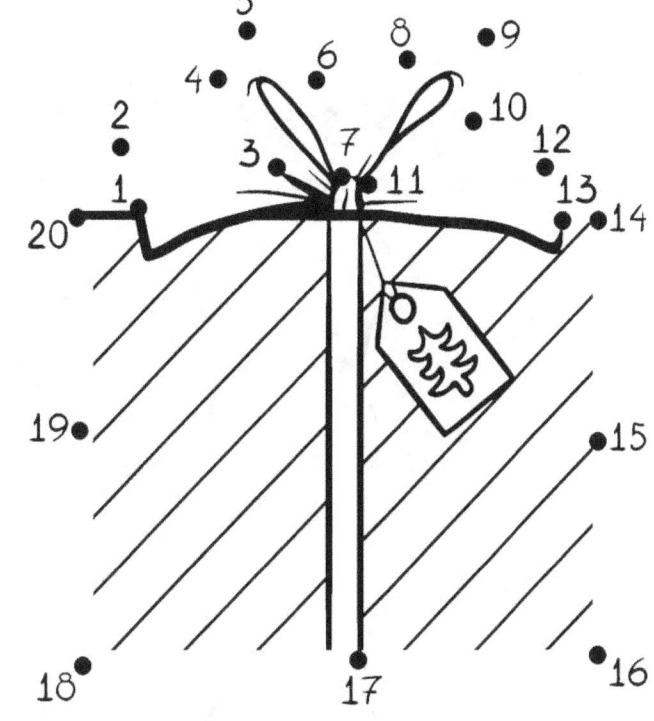

This Book Belongs To:

_____

_____

_____

_____

Test Your Color

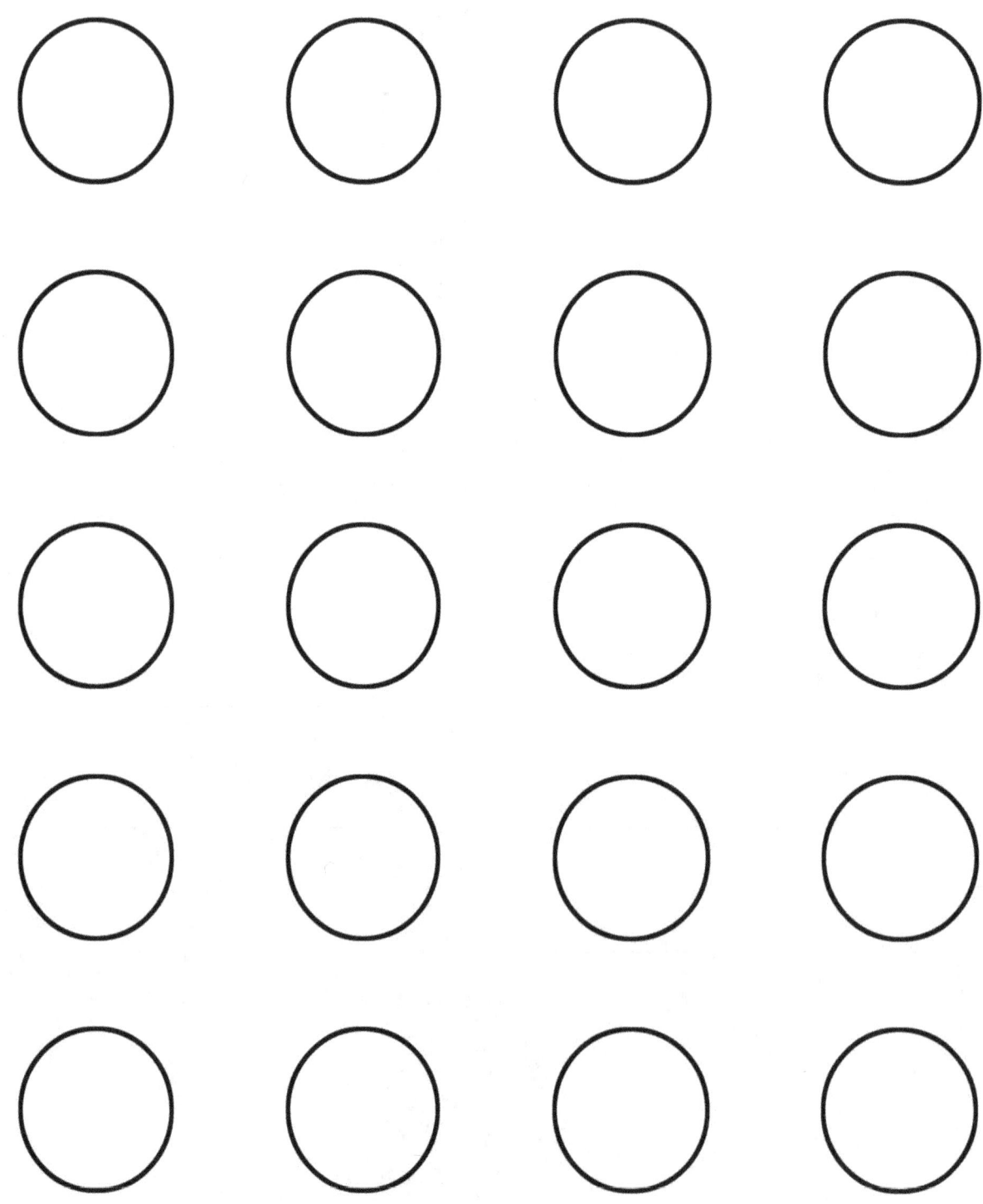

Continue to join the dots you have connected all the numbered dots.
Then, color the picture!

Test Your Color

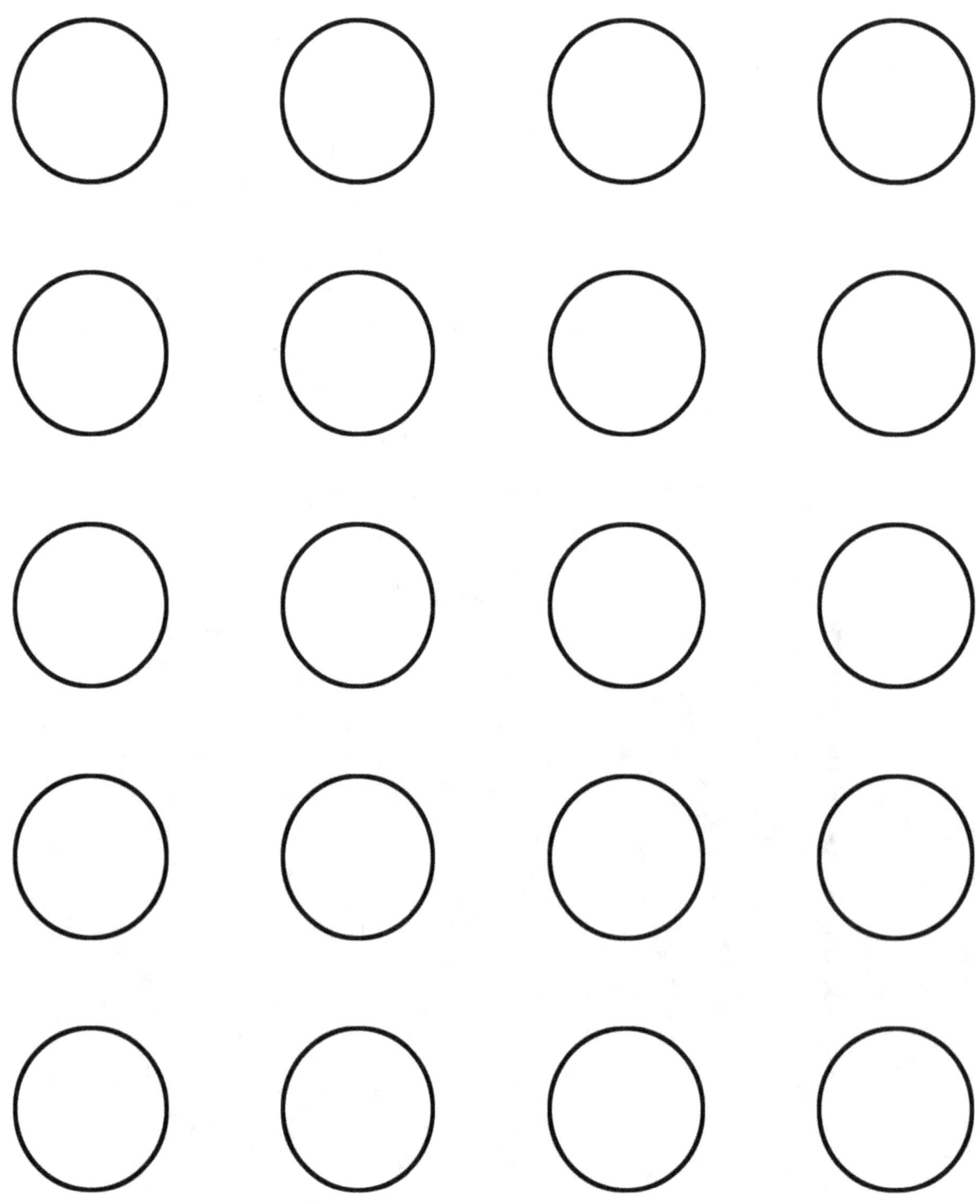

Continue to join the dots you have connected all the numbered dots. Then, color the picture!

Test Your Color

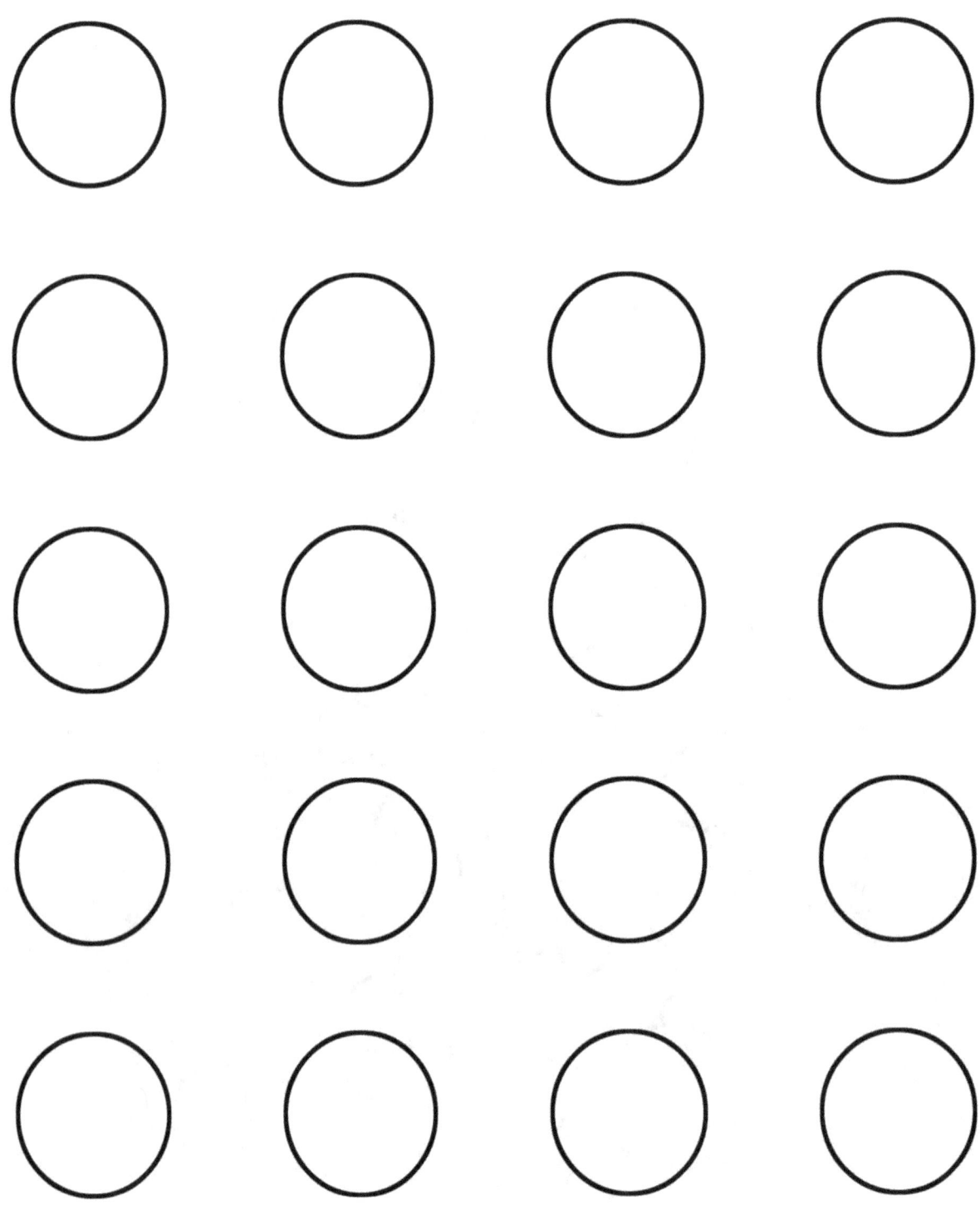

Continue to join the dots you have connected all the numbered dots. Then, color the picture!

Test Your Color

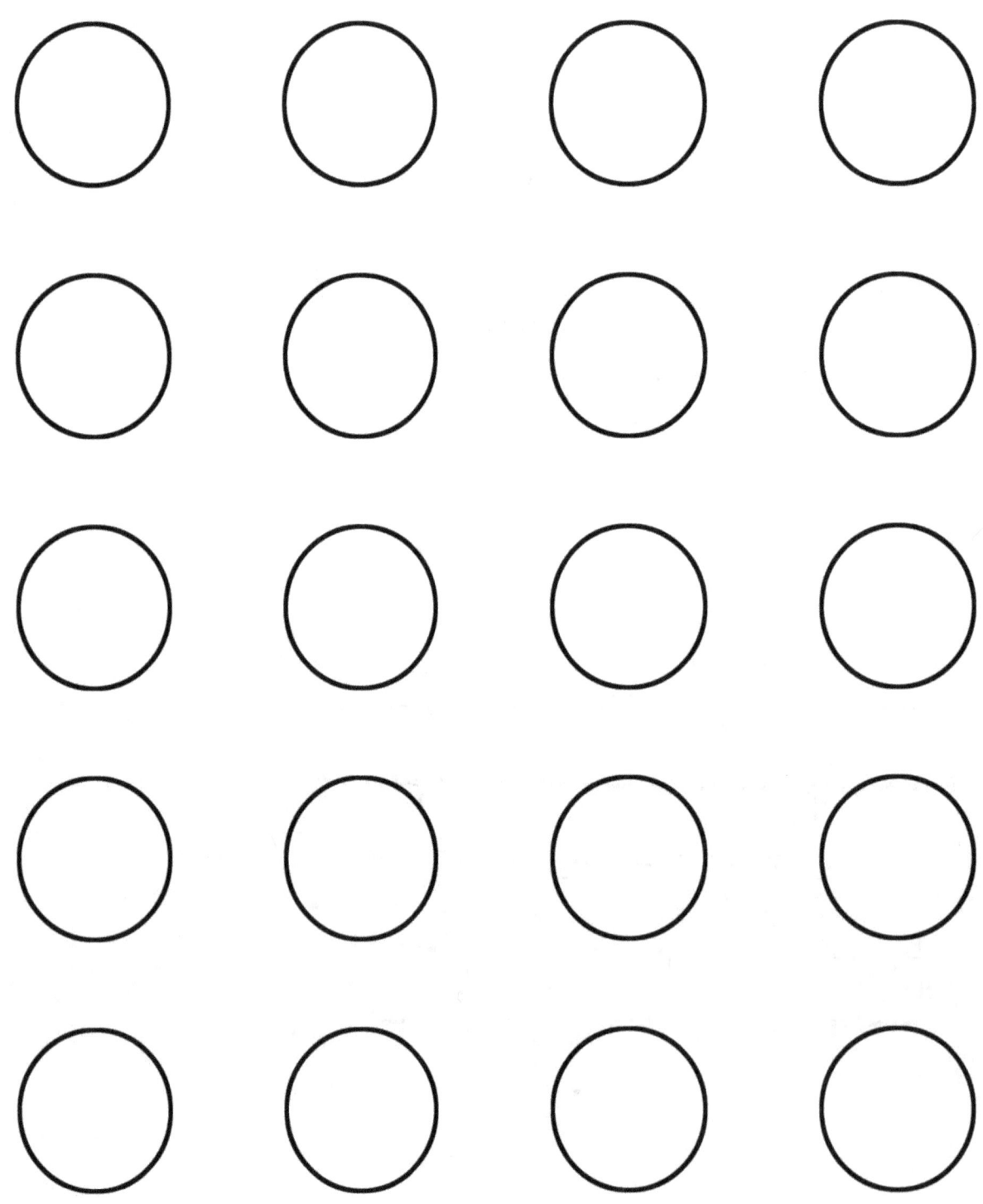

Continue to join the dots you have connected all the numbered dots.
Then, color the picture!

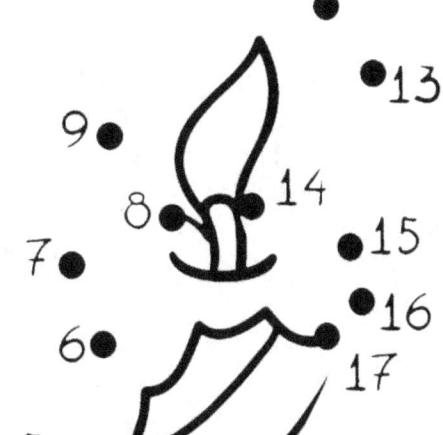
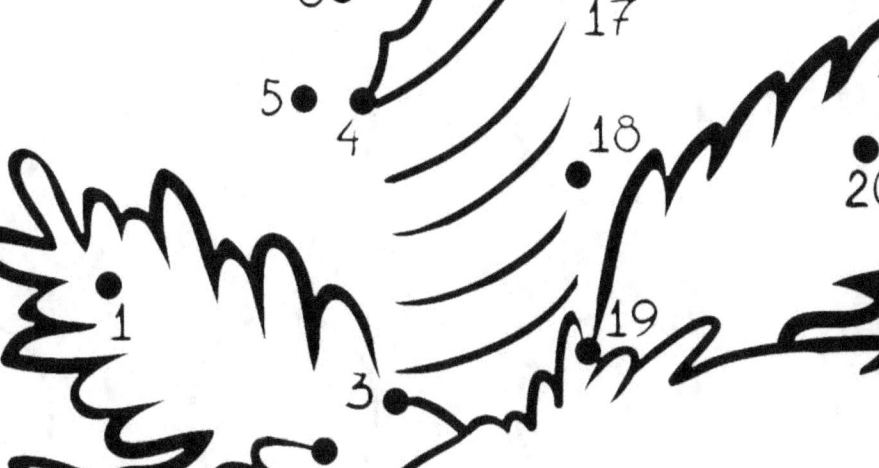

Test Your Color

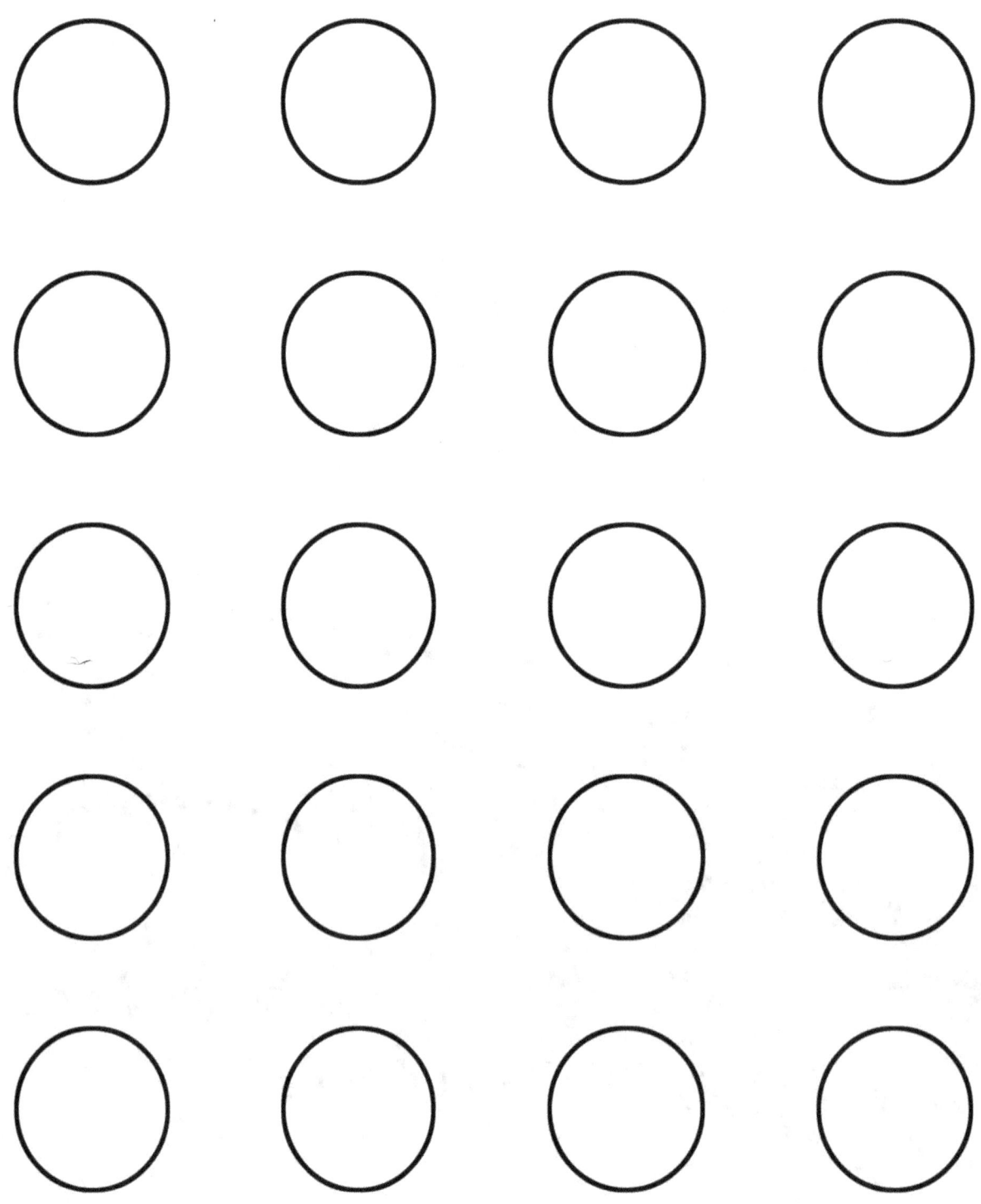

Test Your Color

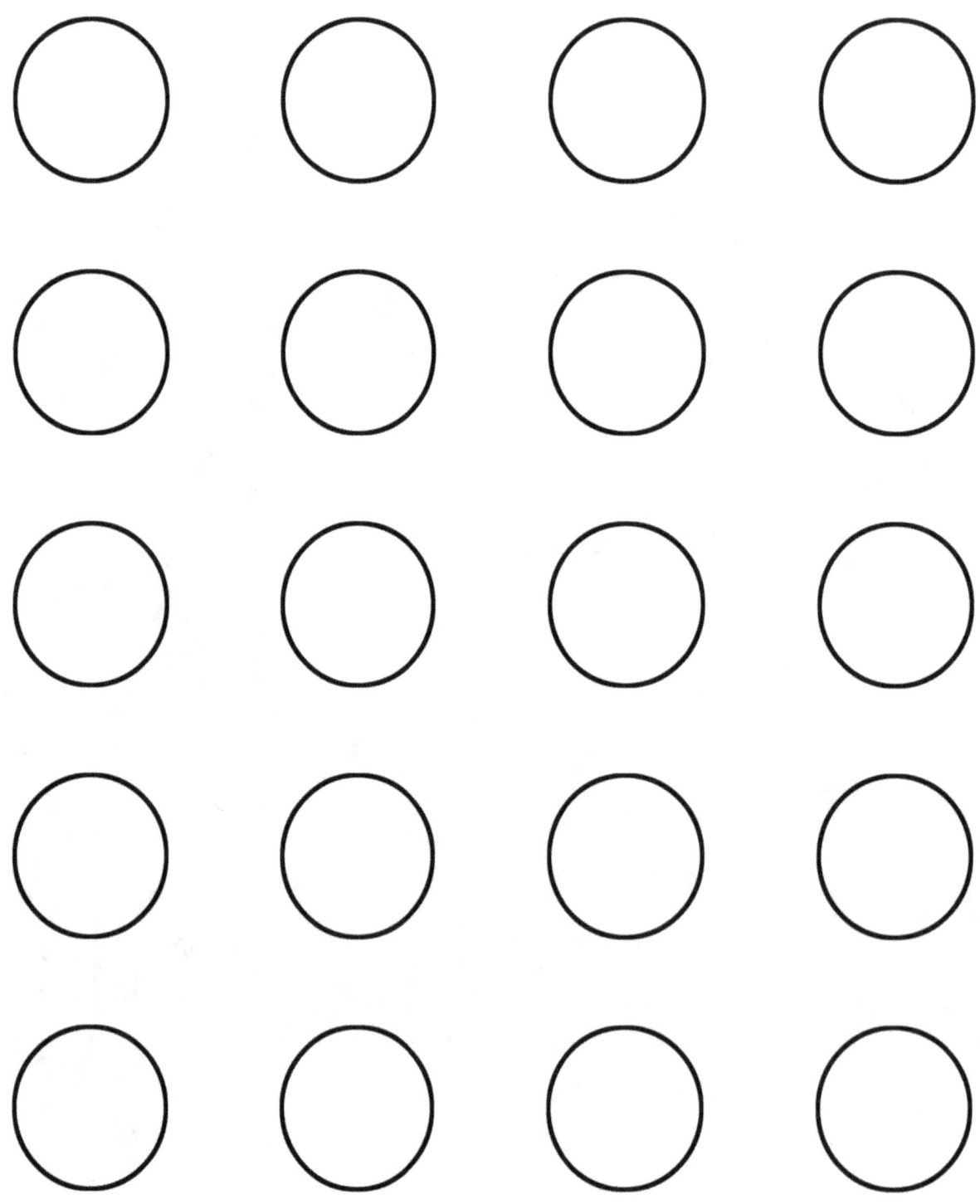

Continue to join the dots you have connected all the numbered dots.
Then, color the picture!

Test Your Color

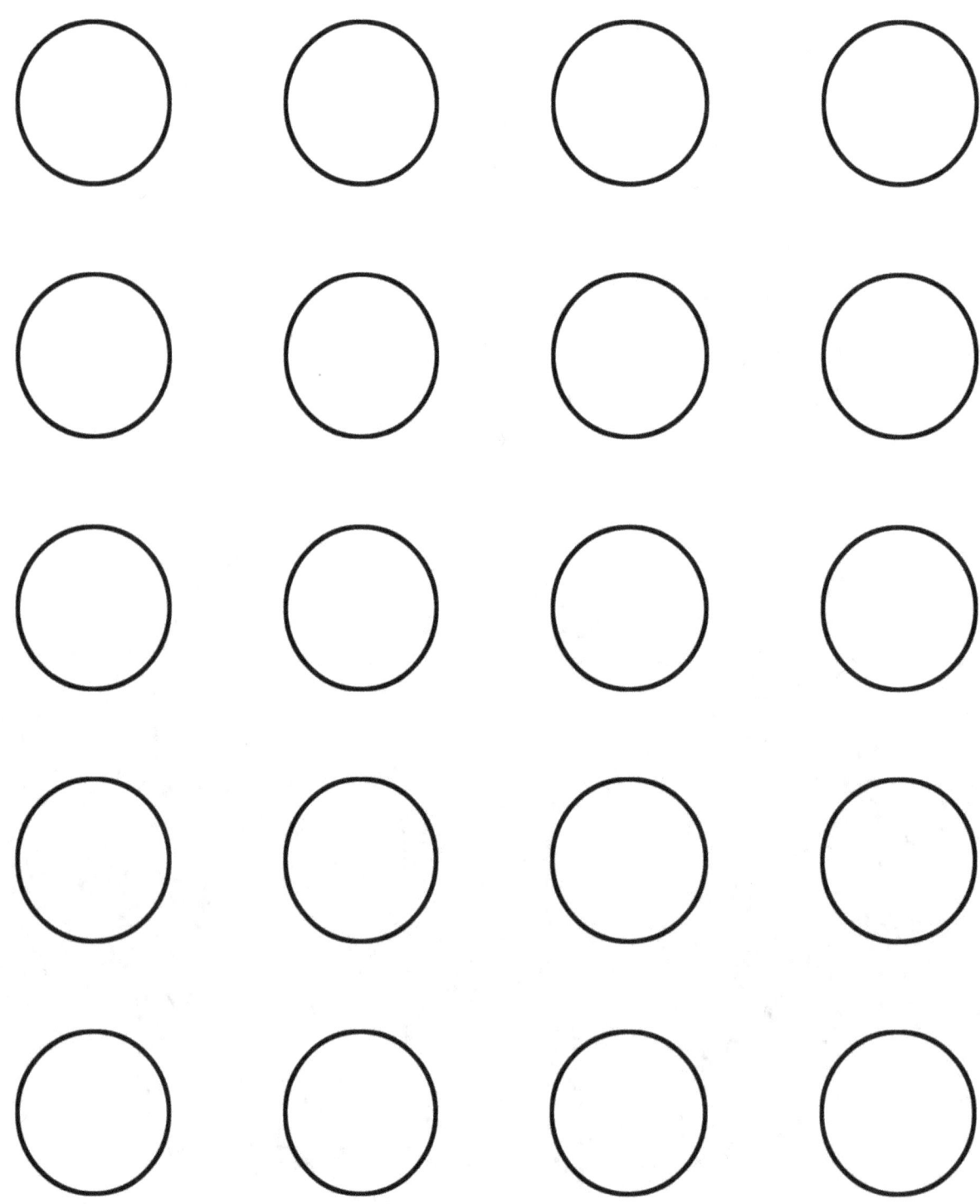

Continue to join the dots you have connected all the numbered dots.
Then, color the picture!

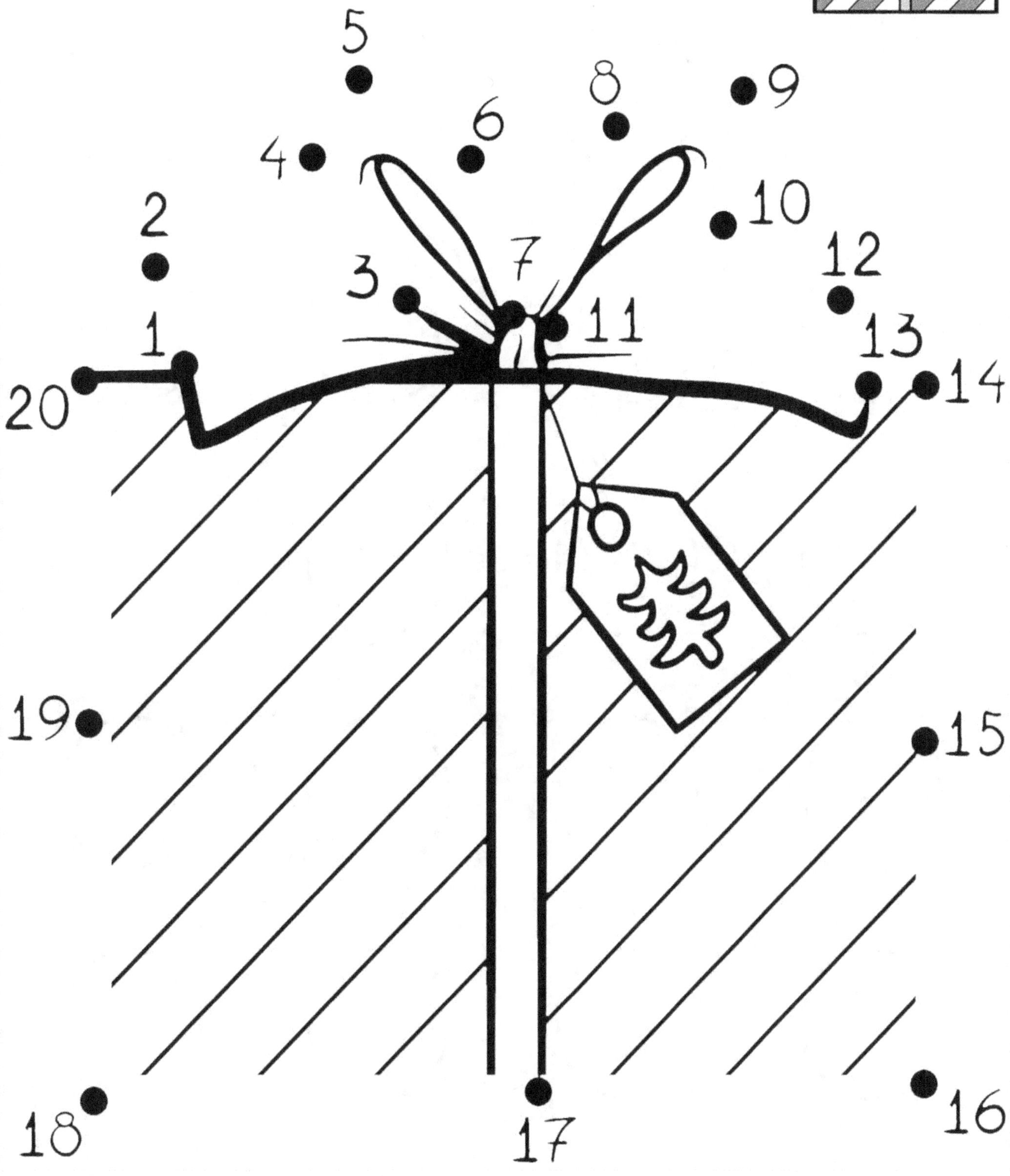

Test Your Color

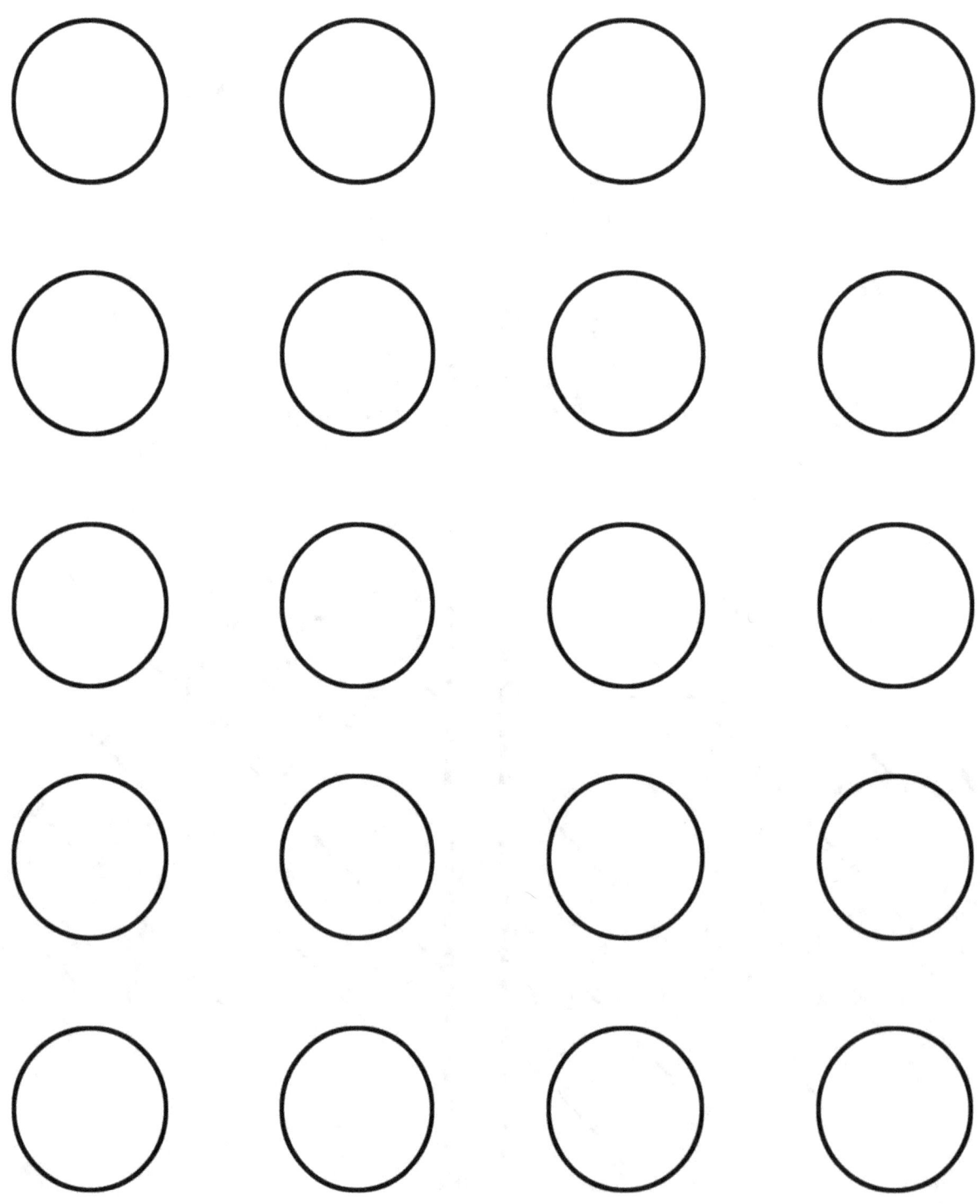

Continue to join the dots you have connected all the numbered dots. Then, color the picture!

Test Your Color

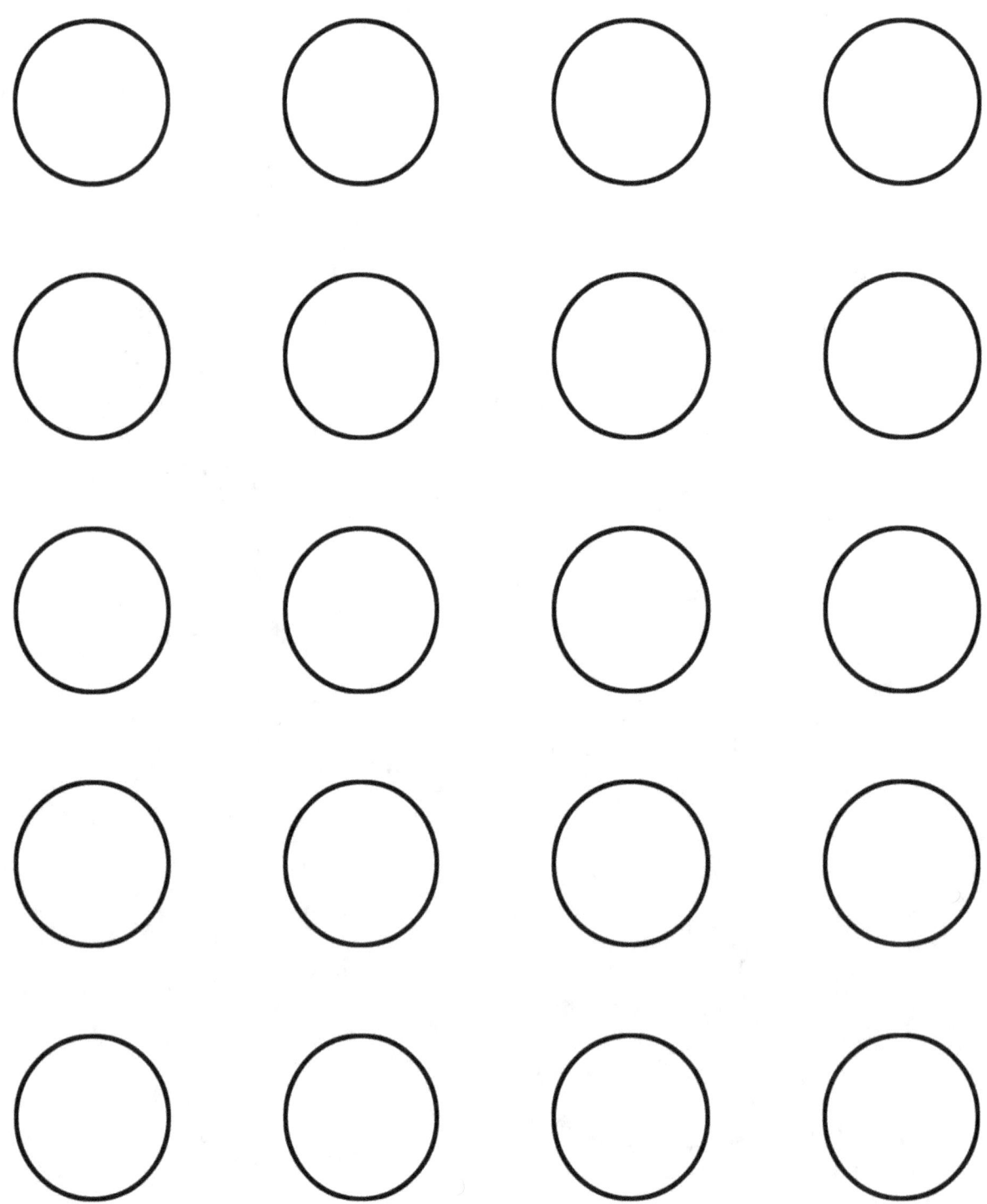

Continue to join the dots you have connected all the numbered dots. Then, color the picture!

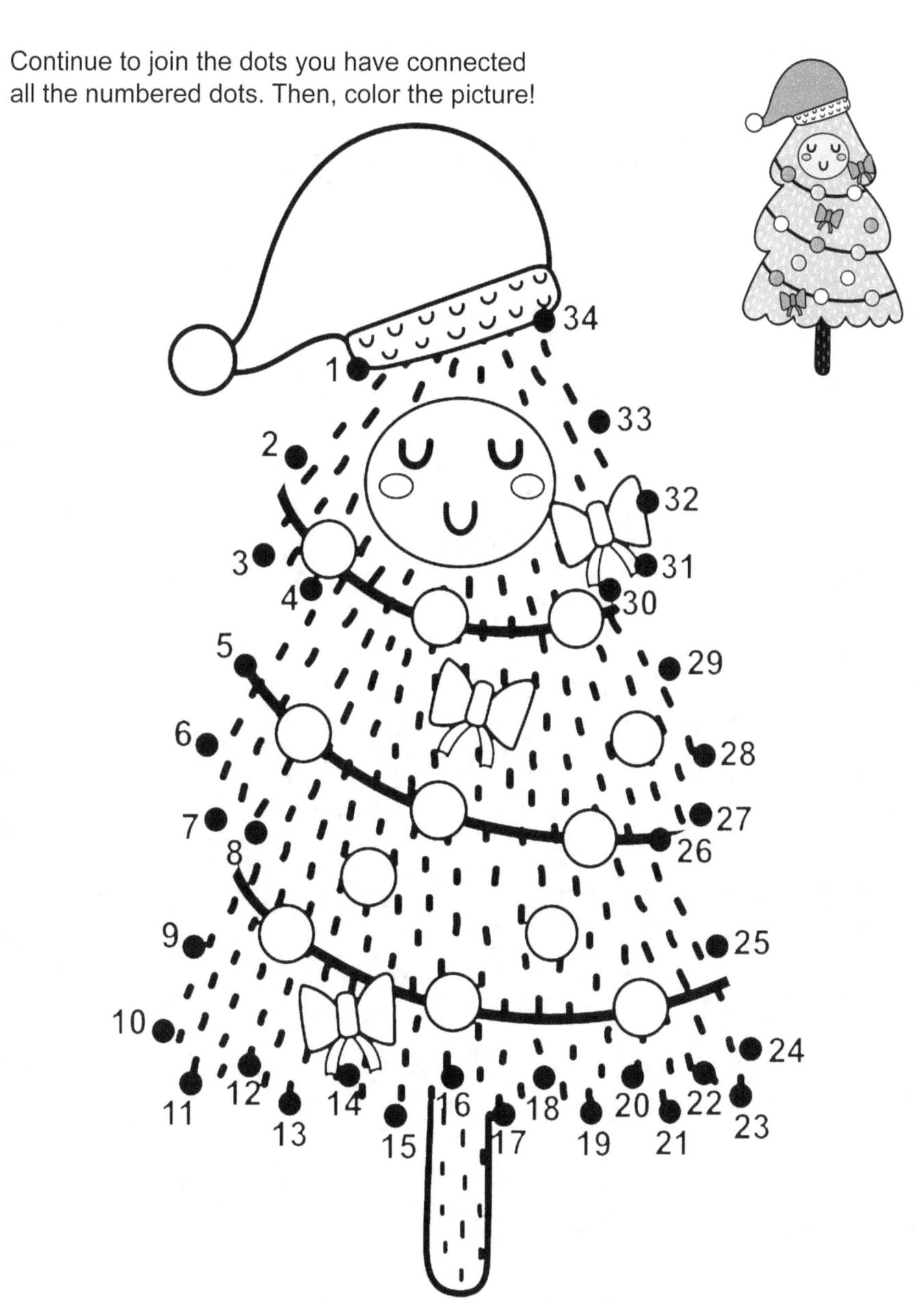

Test Your Color

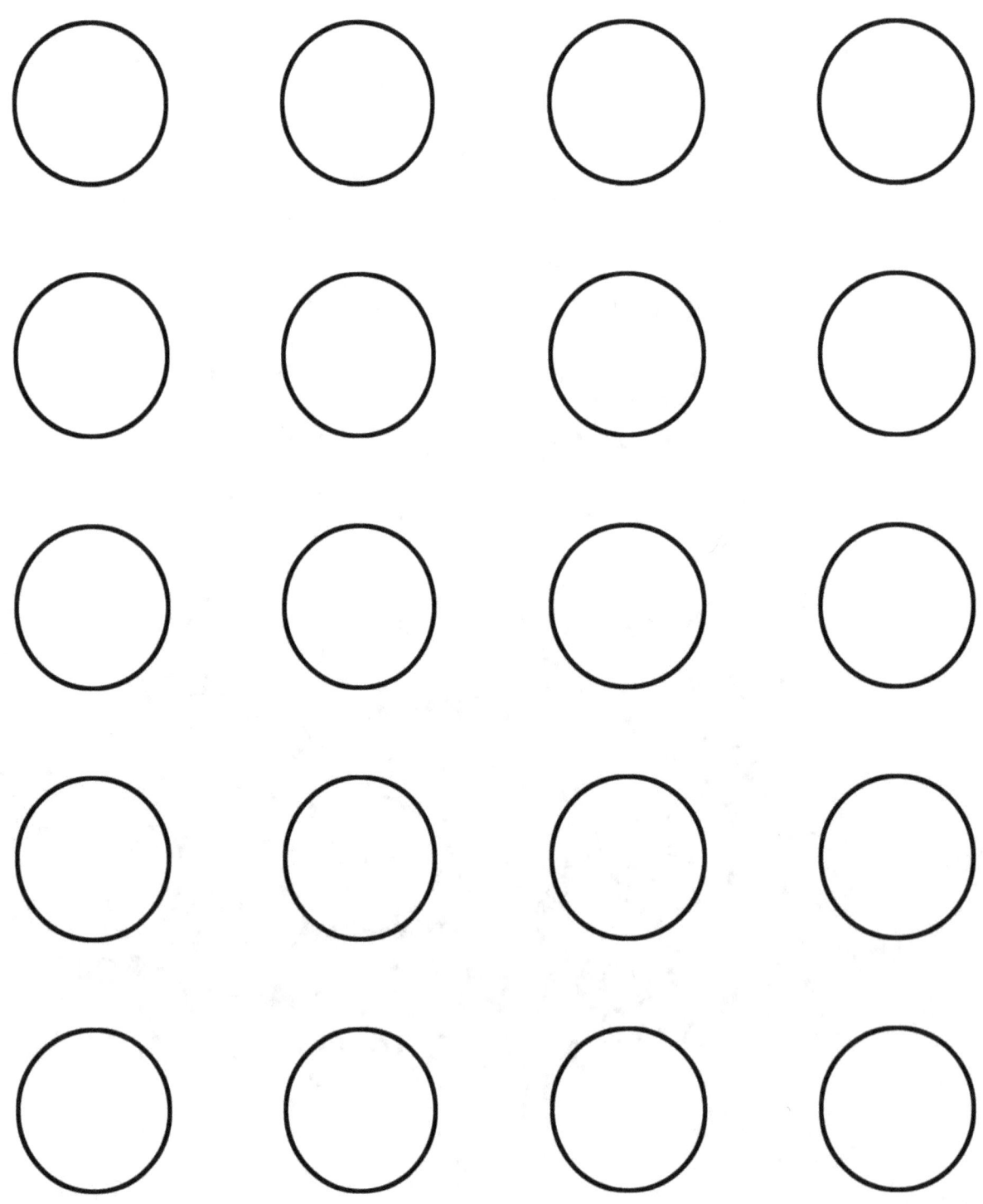

Continue to join the dots you have connected all the numbered dots.
Then, color the picture!

Test Your Color

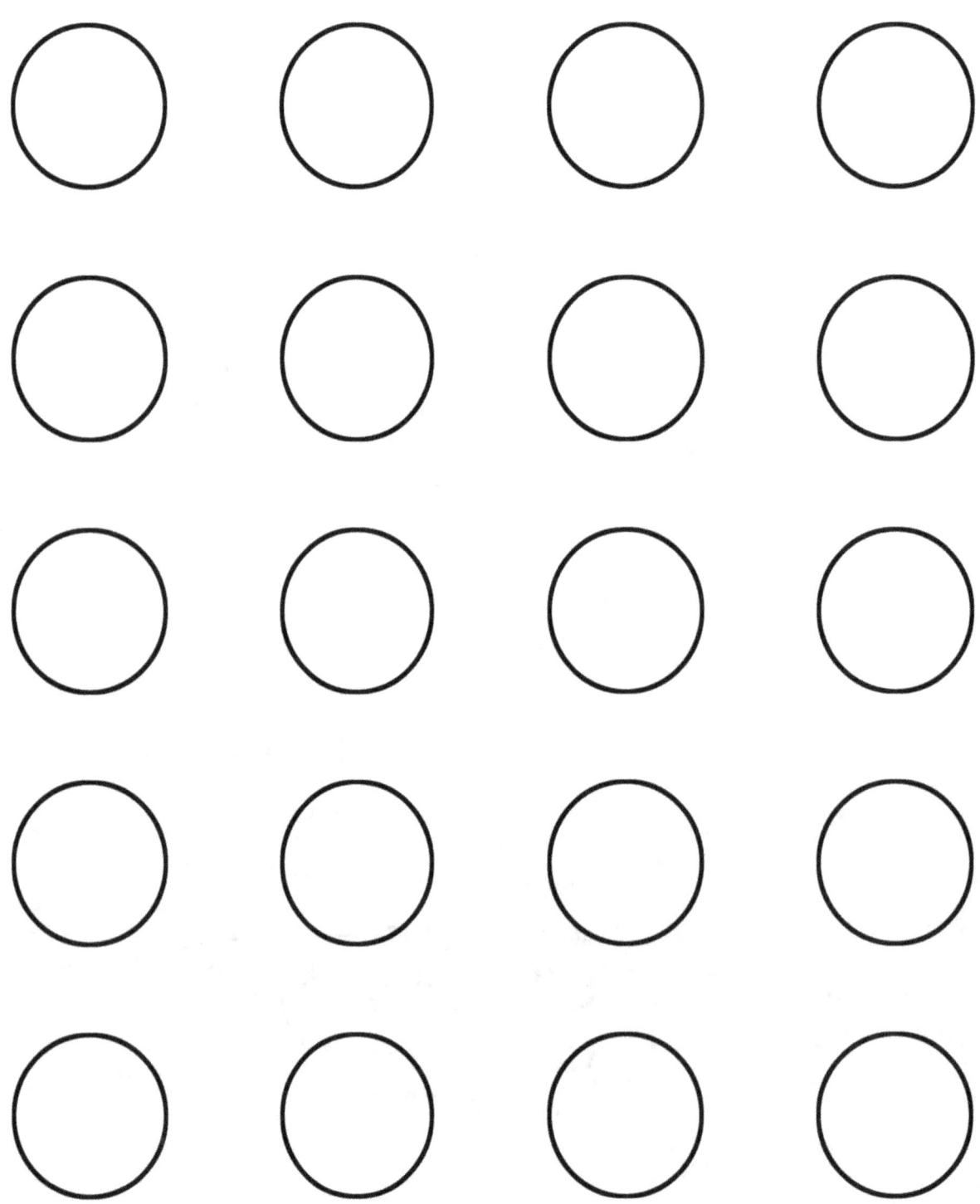

Continue to join the dots you have connected all the numbered dots.
Then, color the picture!

Test Your Color

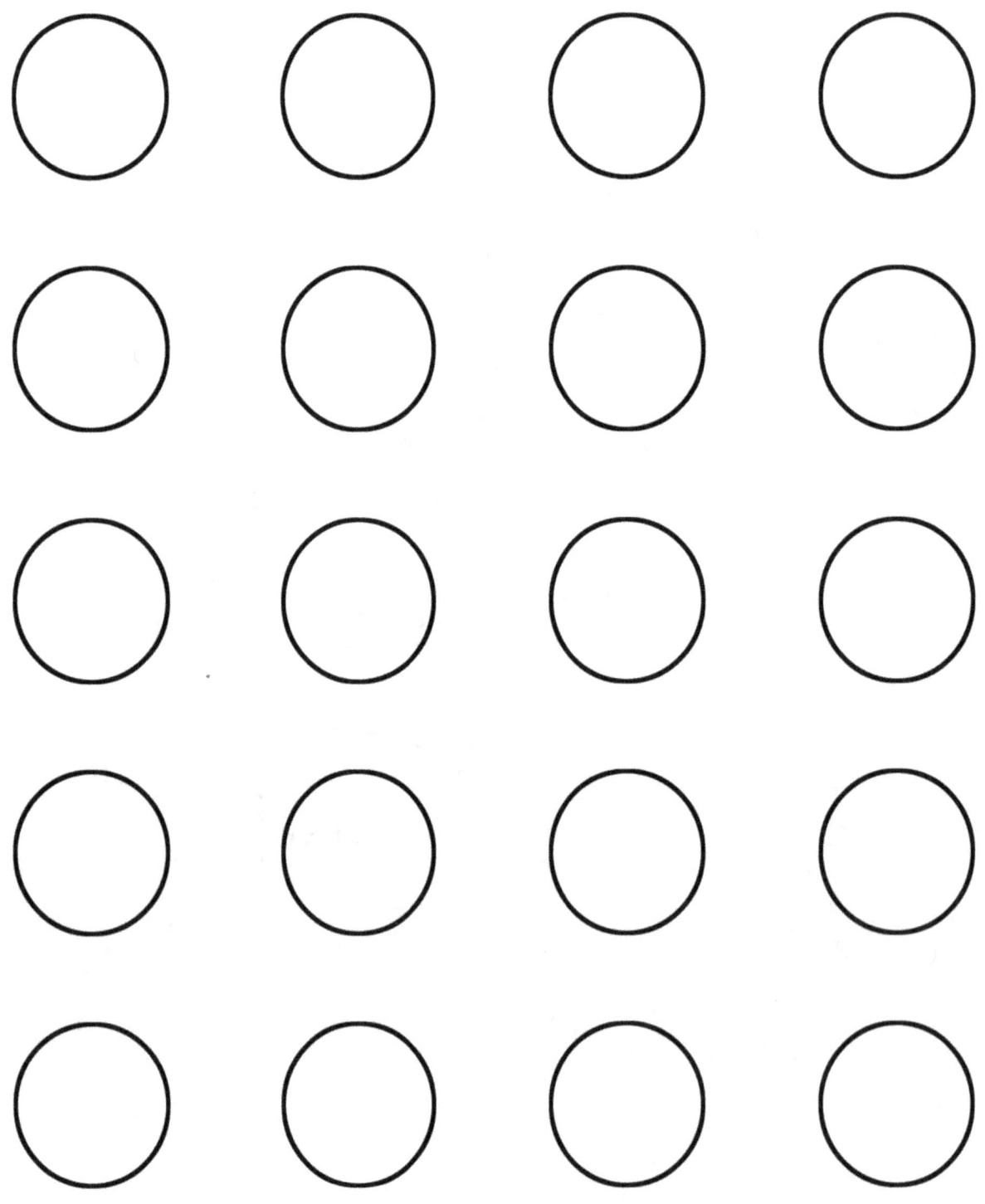

Continue to join the dots you have connected all the numbered dots.
Then, color the picture!

Test Your Color

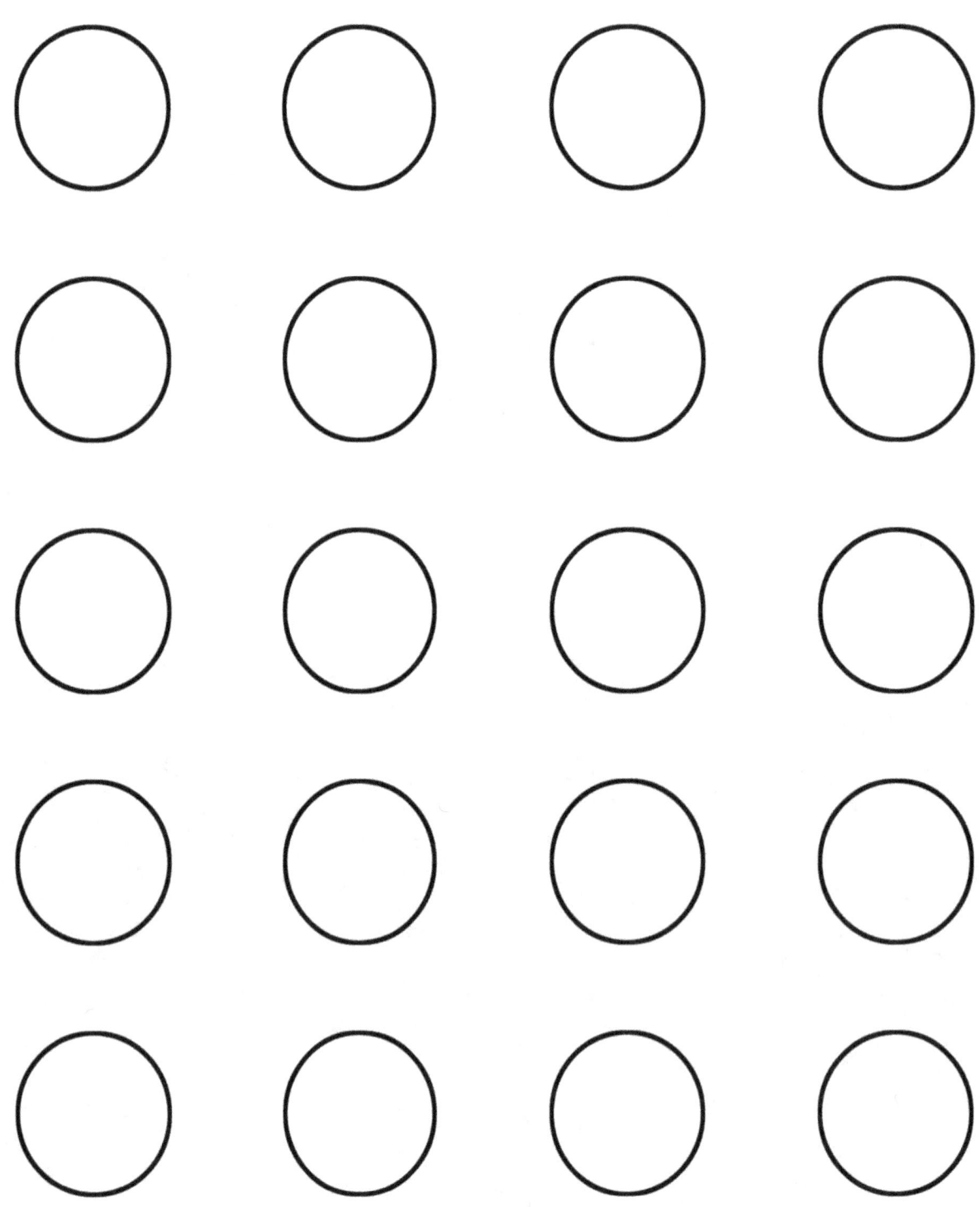

Continue to join the dots you have connected all the numbered dots. Then, color the picture!

Test Your Color

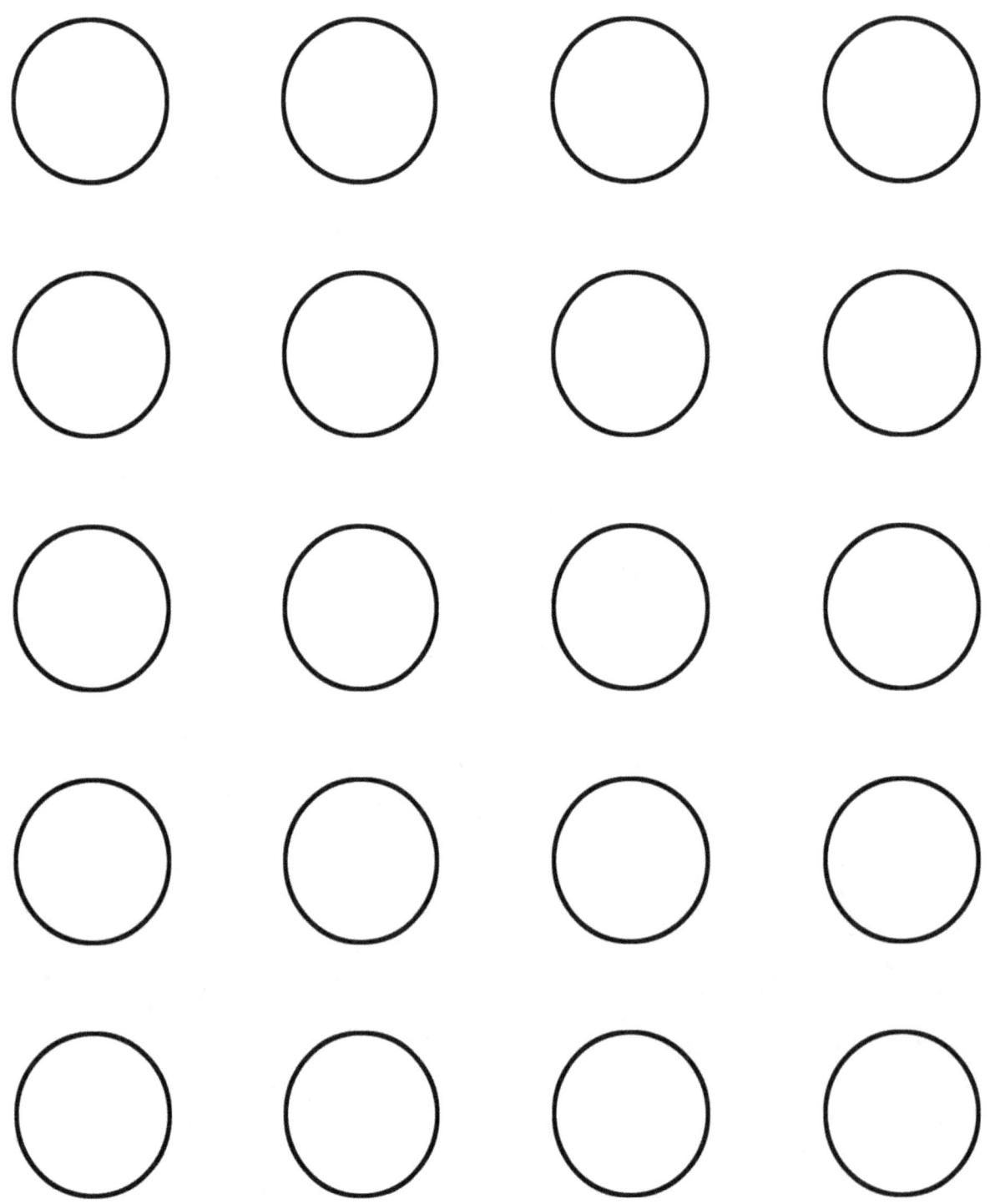

Continue to join the dots you have connected all the numbered dots.
Then, color the picture!

Test Your Color

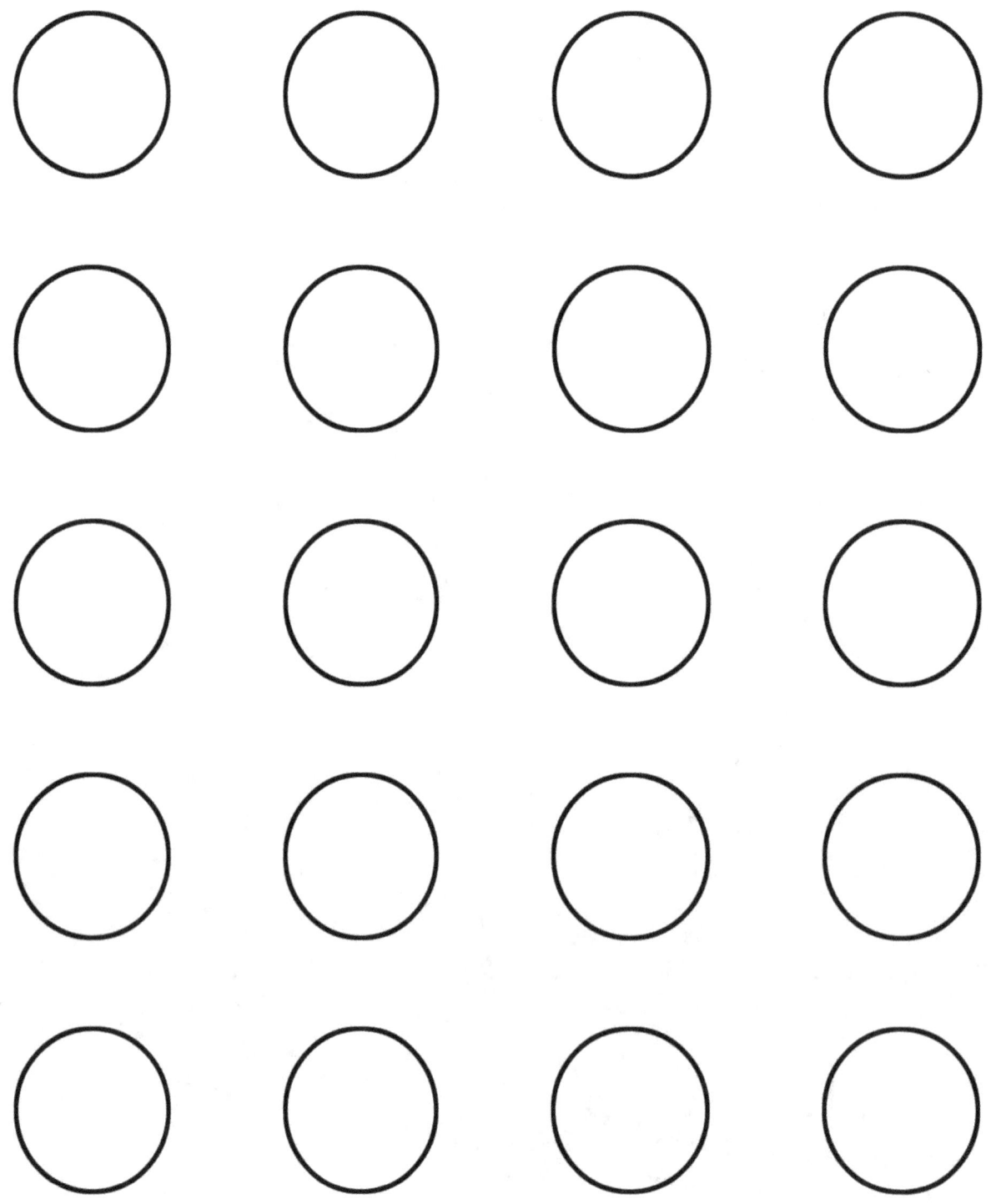

Continue to join the dots you have connected all the numbered dots.
Then, color the picture!

Test Your Color

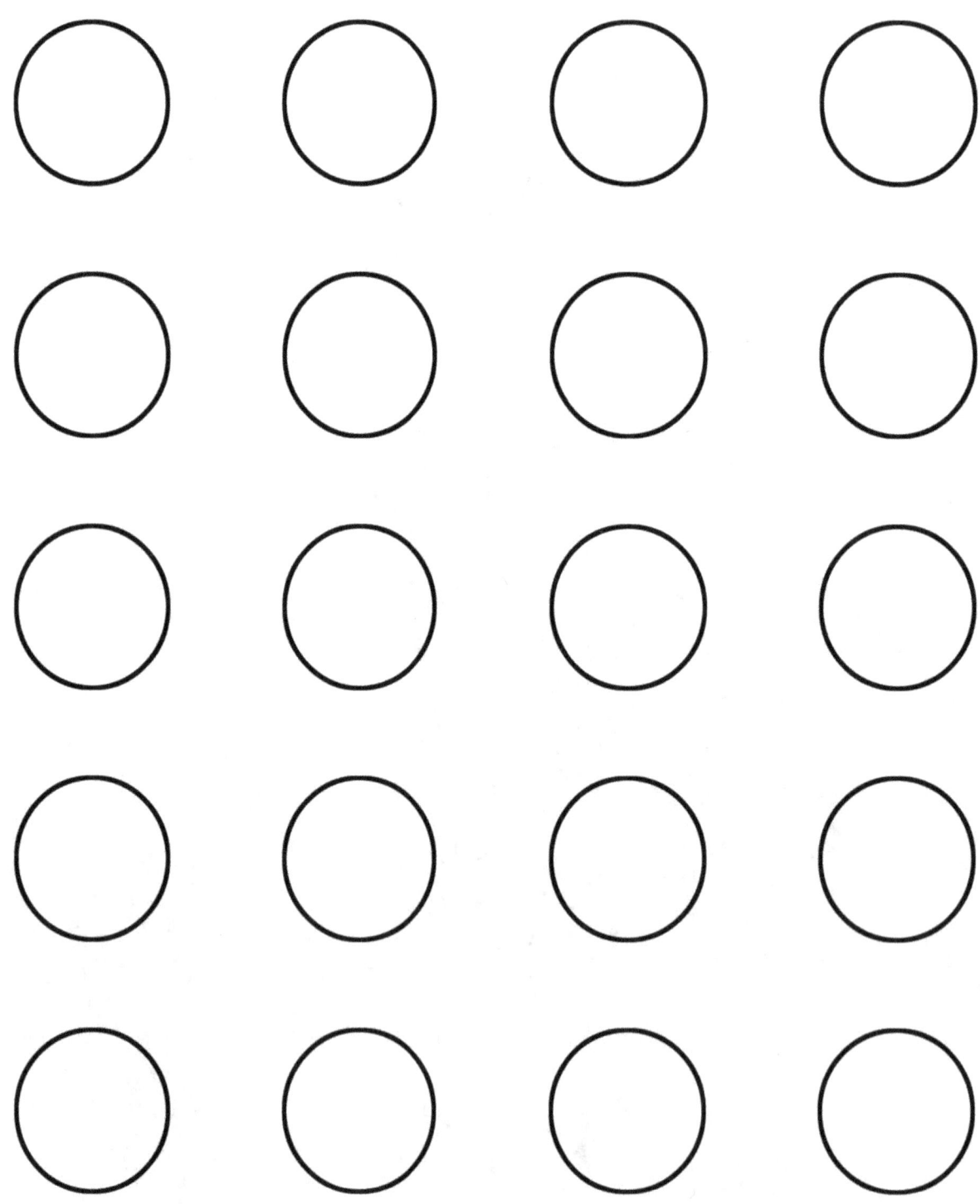

Continue to join the dots you have connected all the numbered dots.
Then, color the picture!

Test Your Color

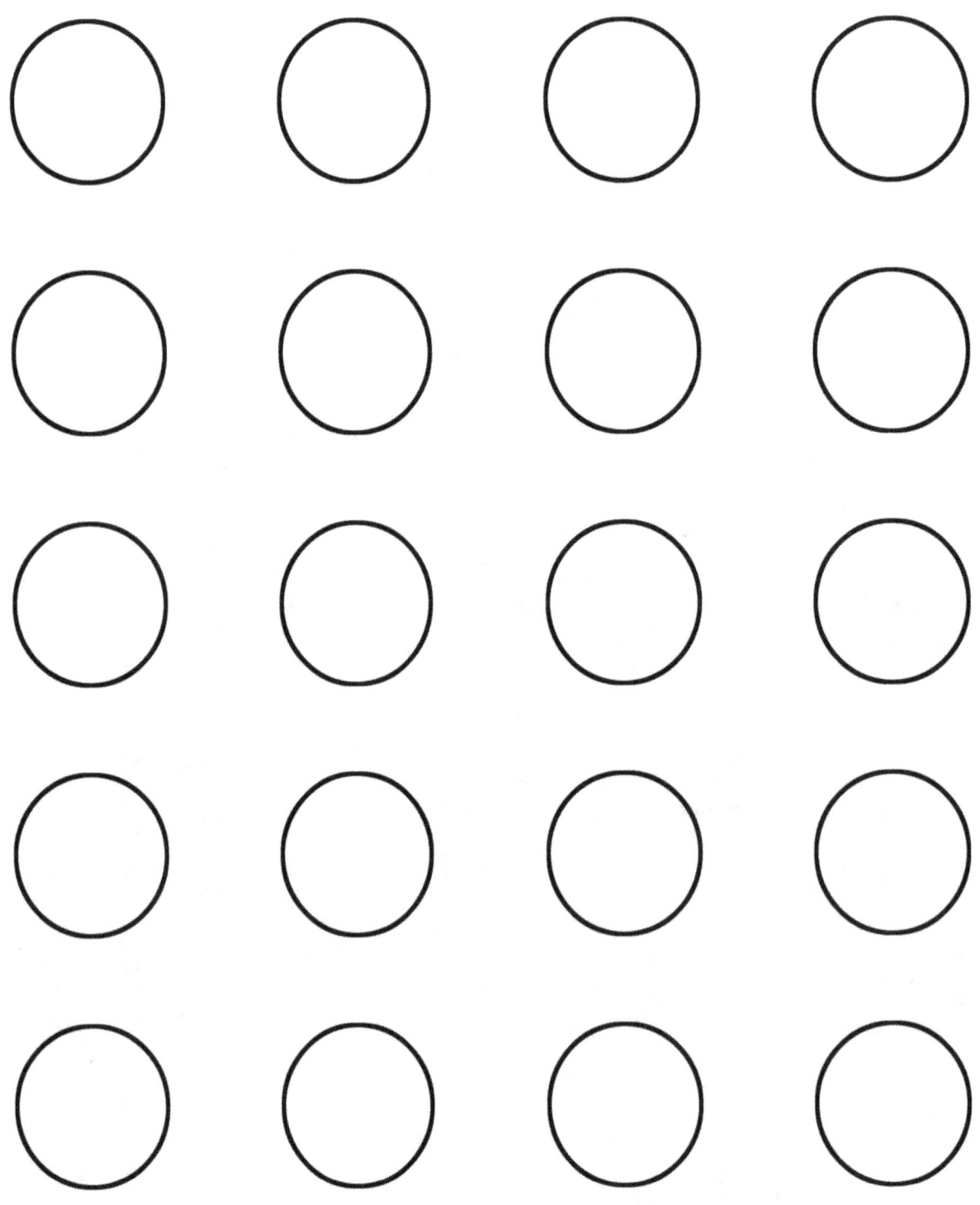

Continue to join the dots you have connected all the numbered dots.
Then, color the picture!

Test Your Color

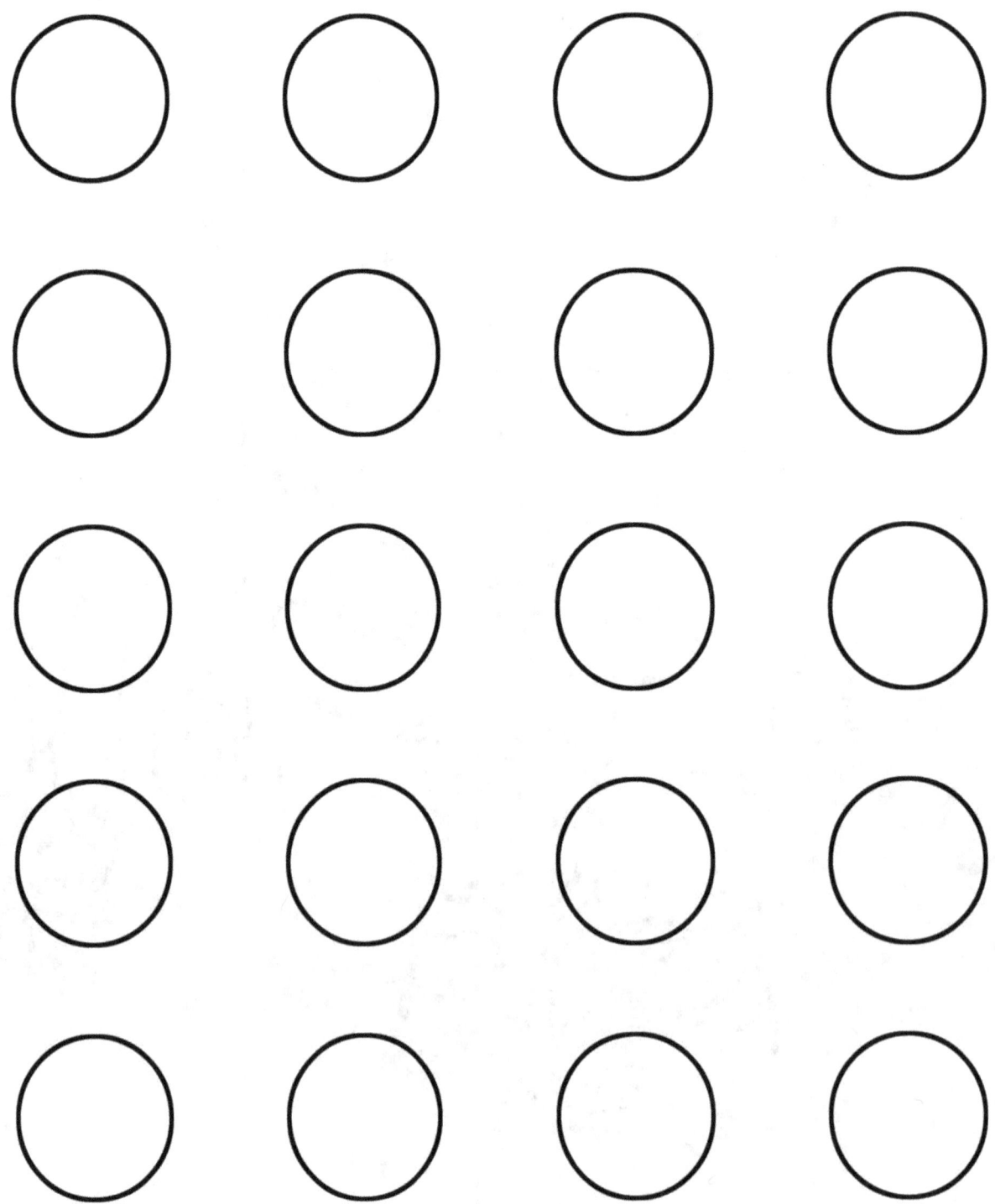

Continue to join the dots you have connected all the numbered dots.
Then, color the picture!

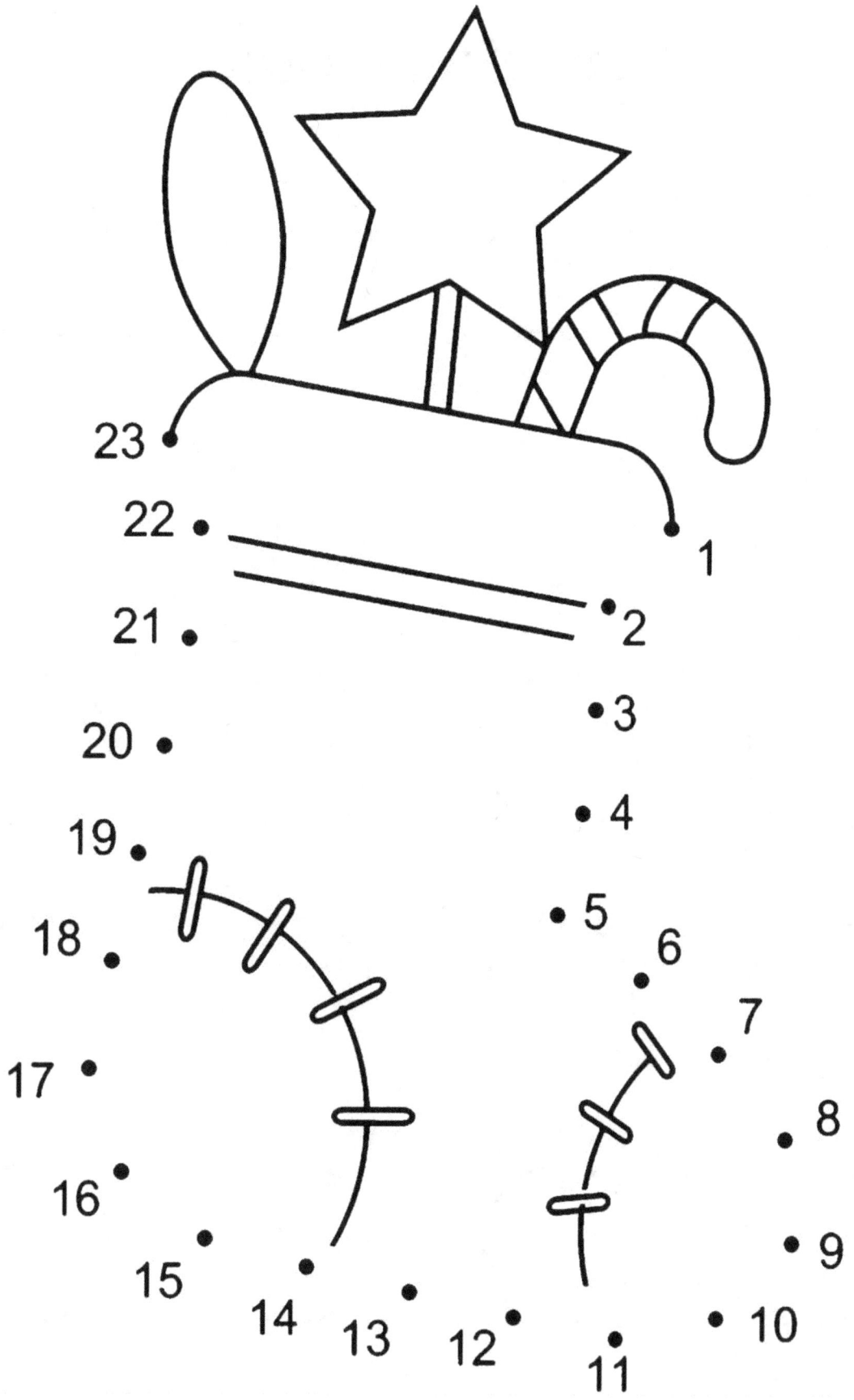

Test Your Color

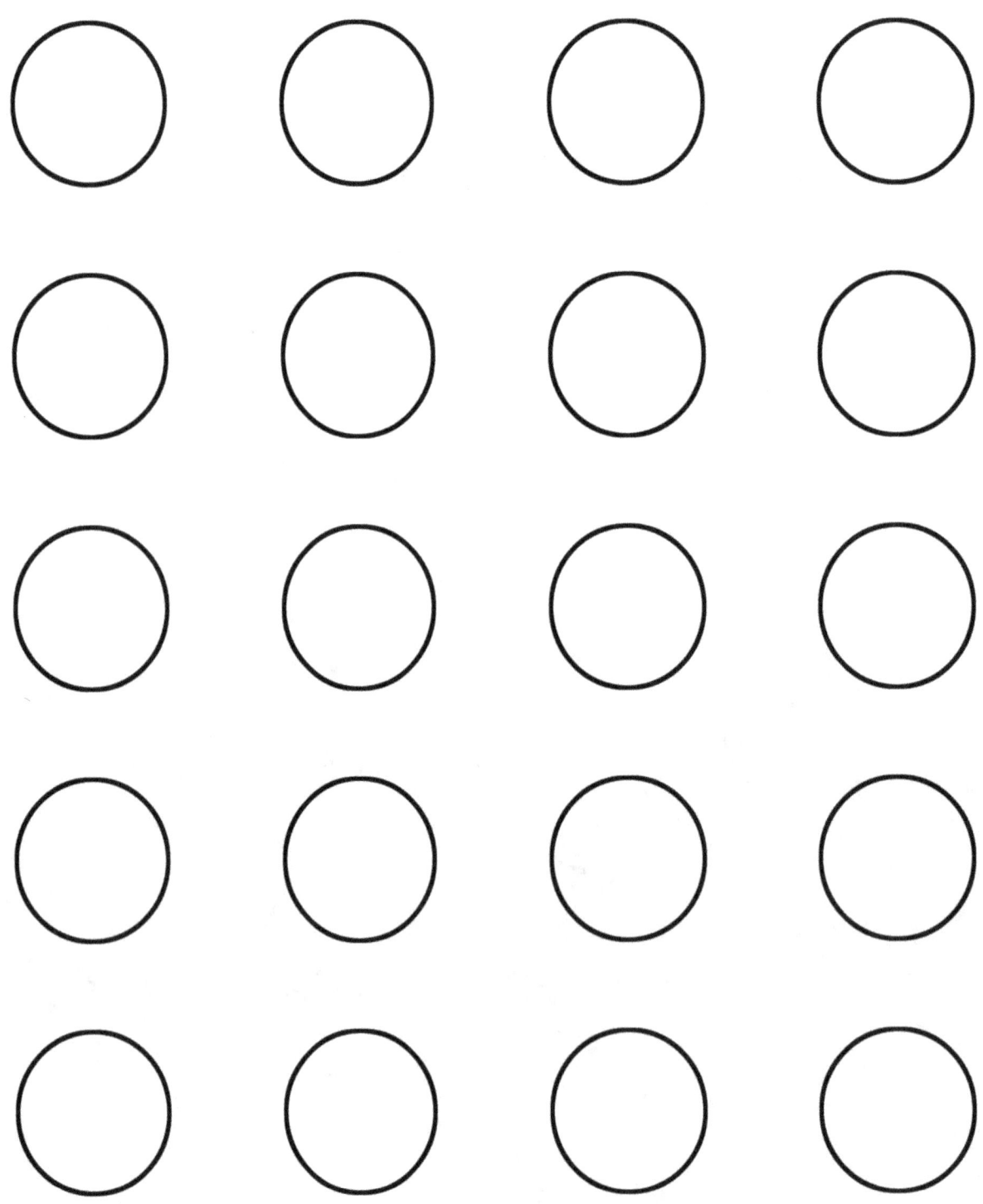

Continue to join the dots you have connected all the numbered dots. Then, color the picture!

Test Your Color

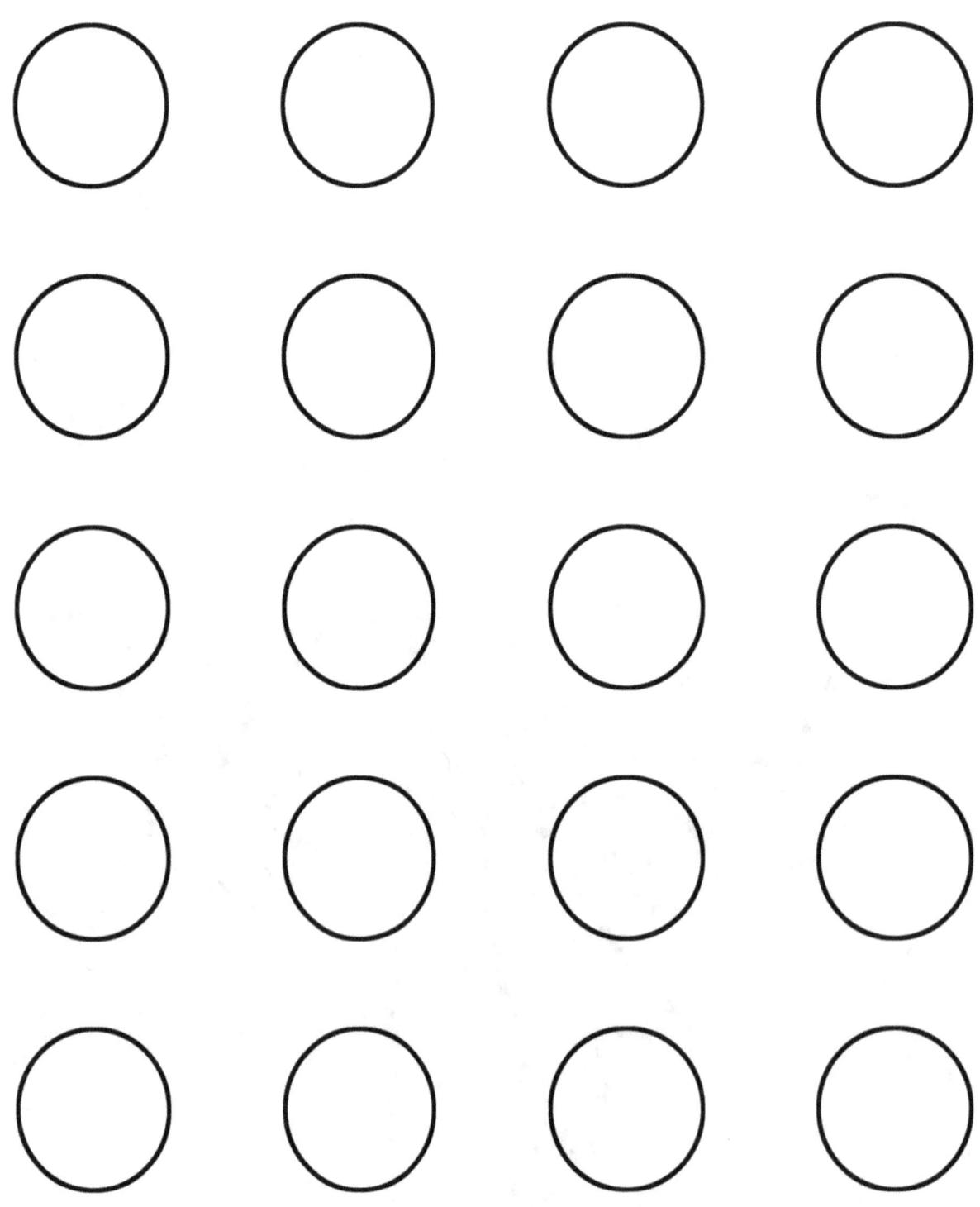

Continue to join the dots you have connected all the numbered dots. Then, color the picture!

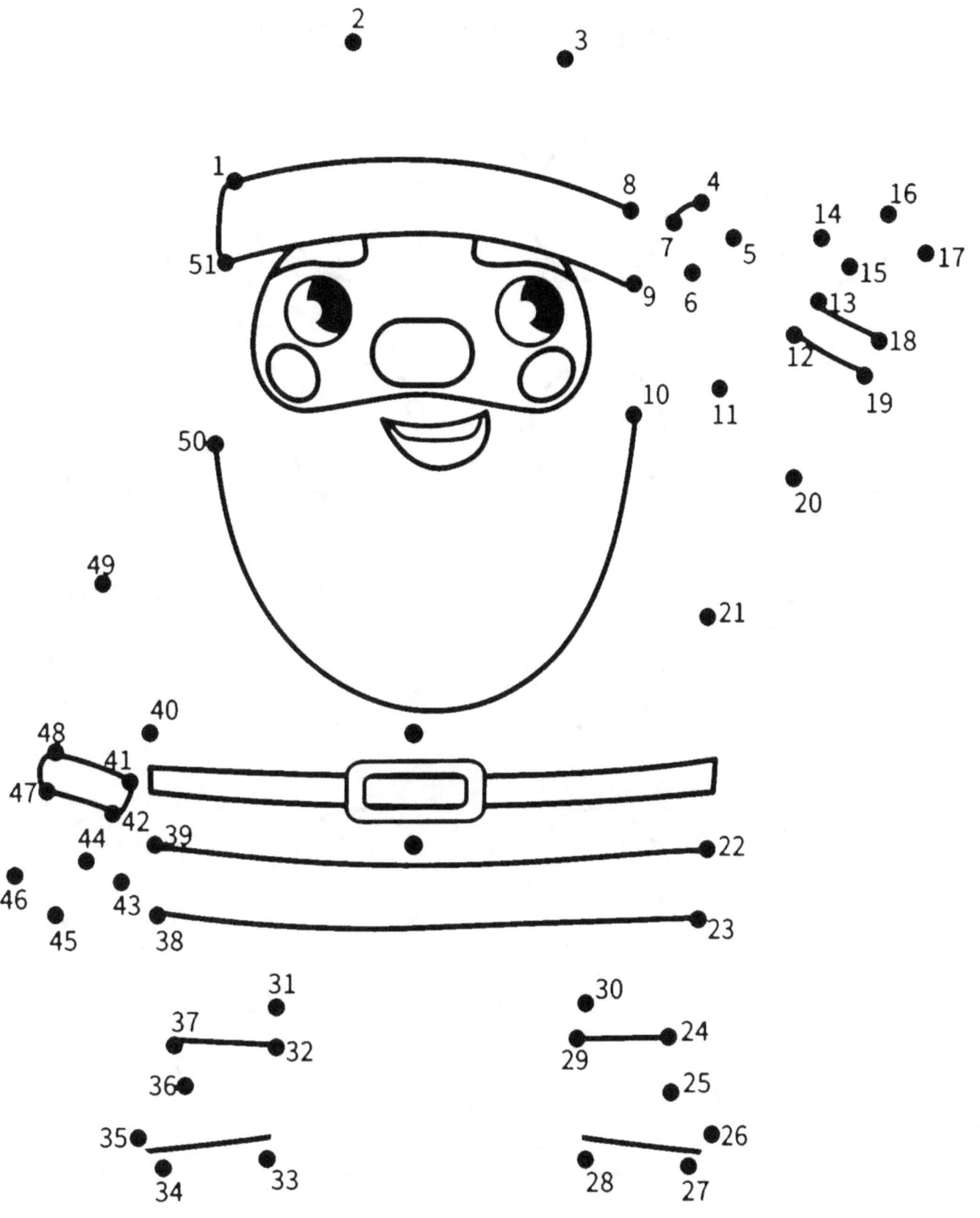

Test Your Color

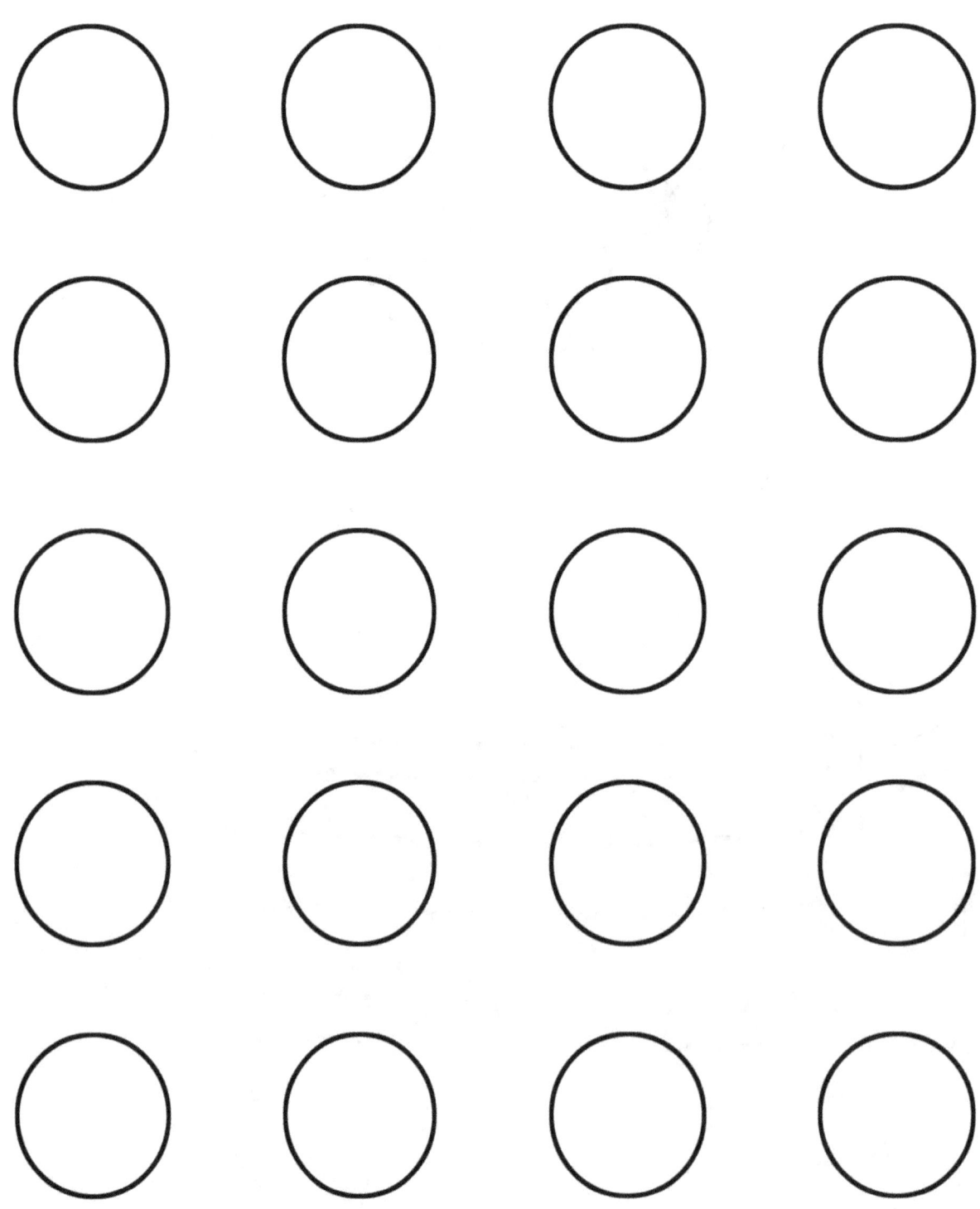

Continue to join the dots you have connected all the numbered dots. Then, color the picture!

Test Your Color

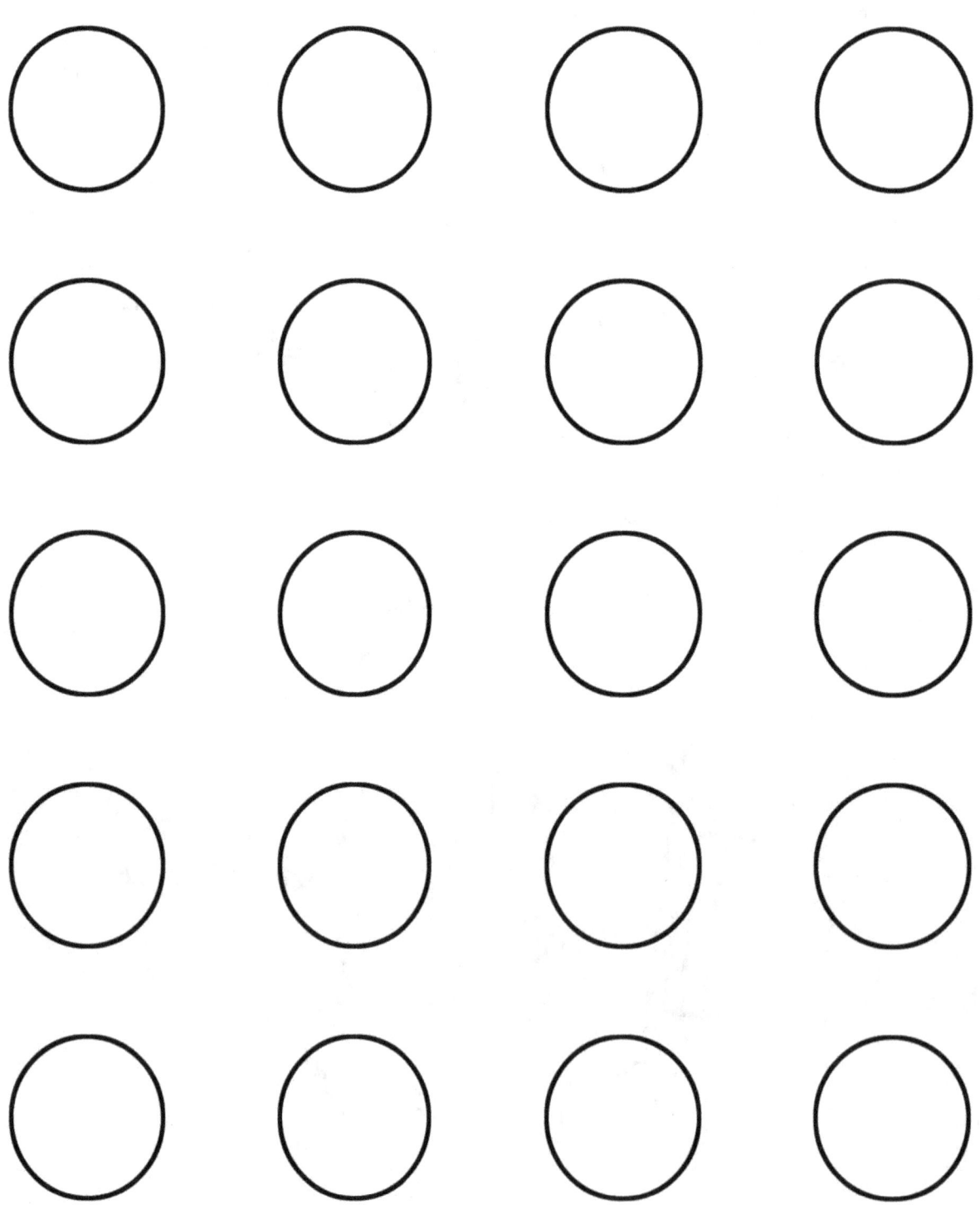

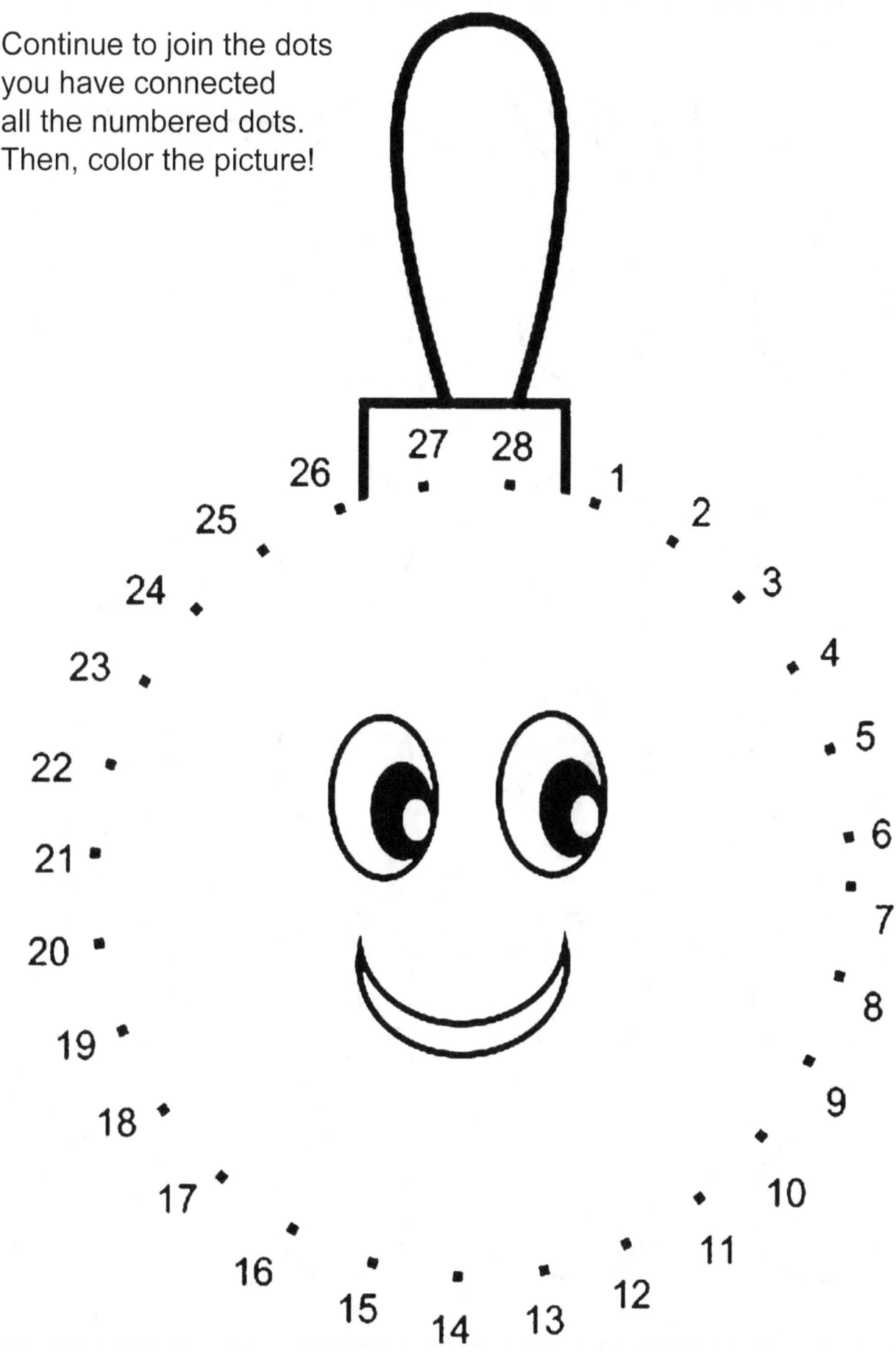

Test Your Color

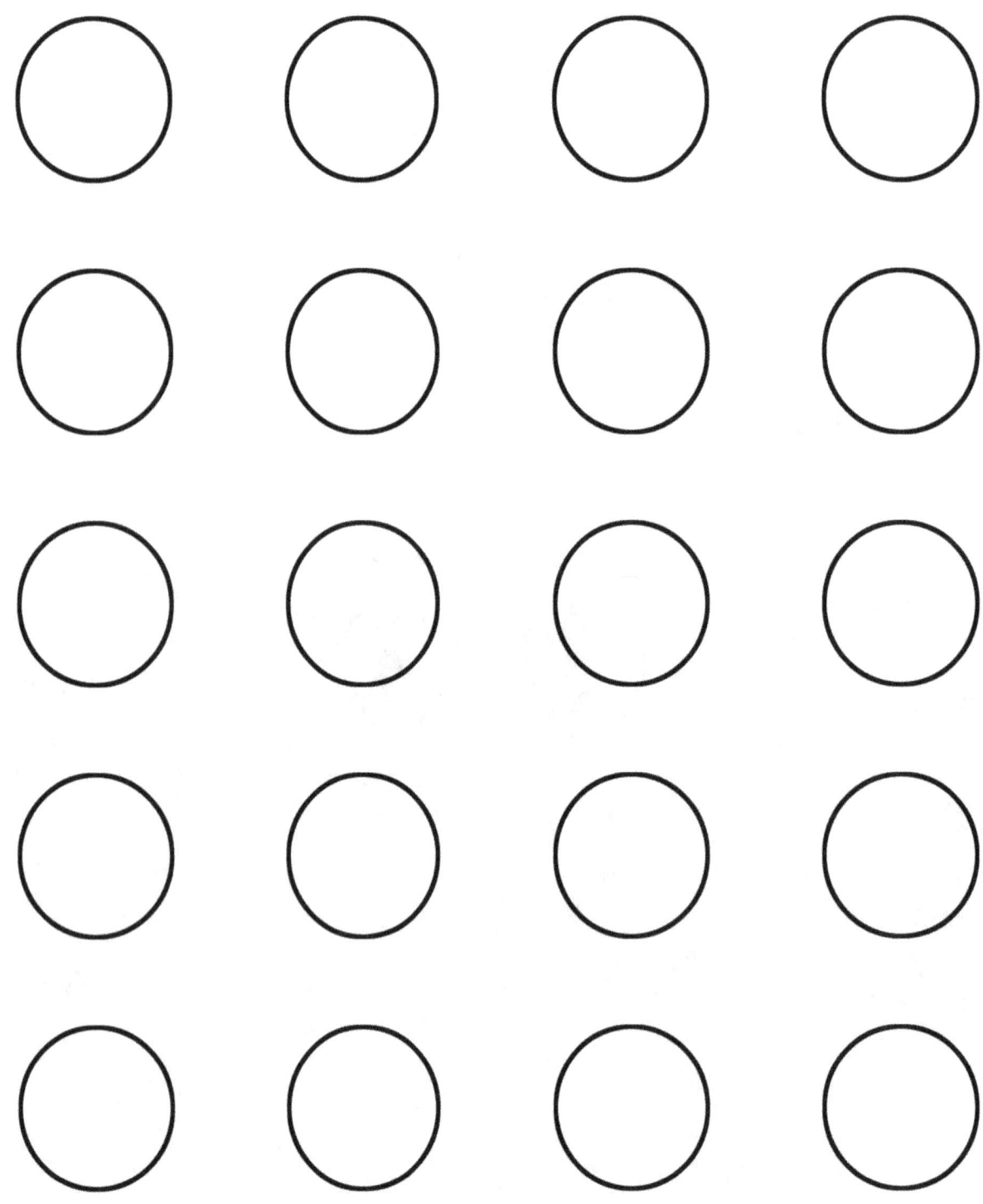

Continue to join the dots you have connected all the numbered dots. Then, color the picture!

Test Your Color

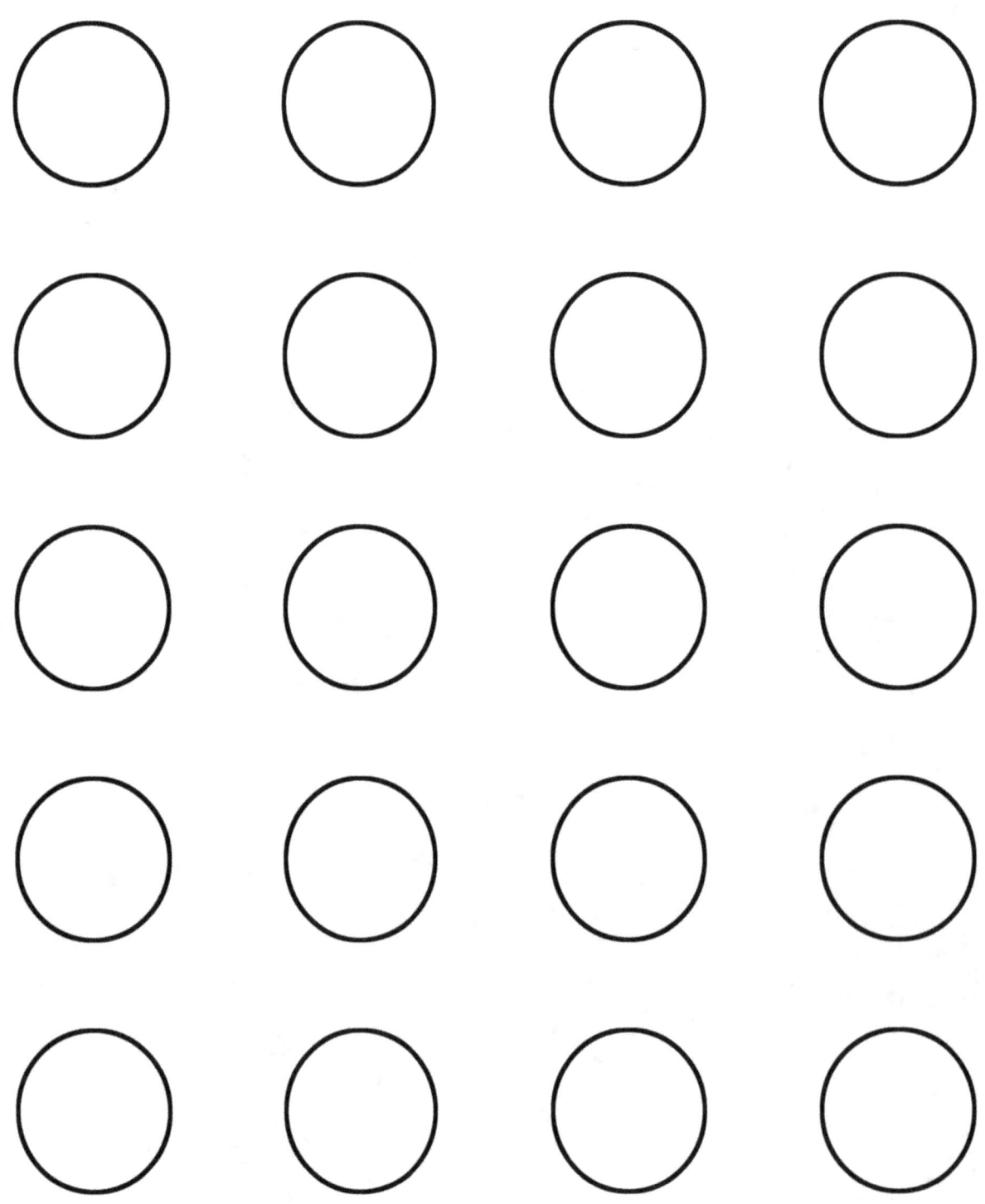

Continue to join the dots you have connected all the numbered dots. Then, color the picture!

Test Your Color

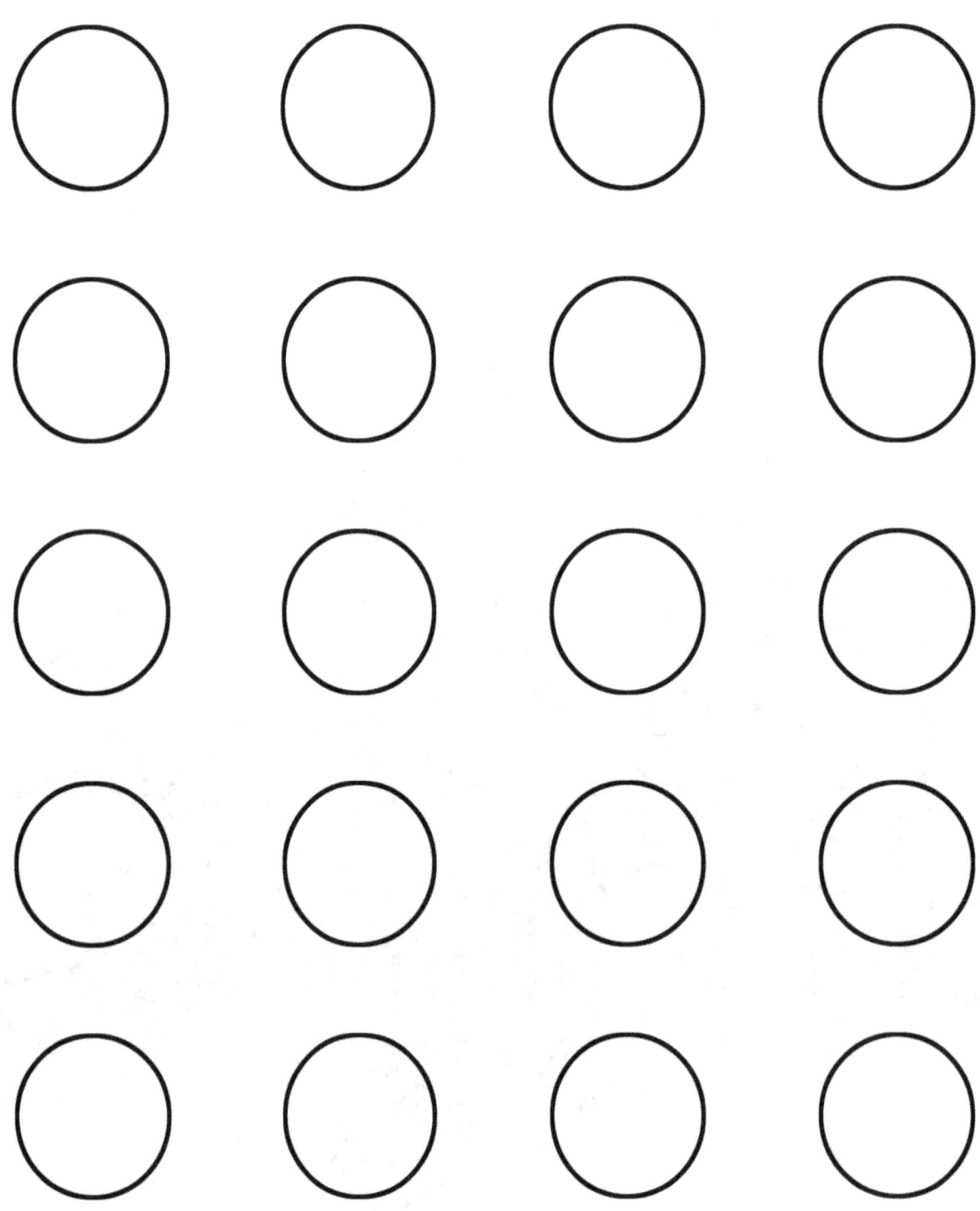

Continue to join the dots you have connected all the numbered dots. Then, color the picture!

Test Your Color

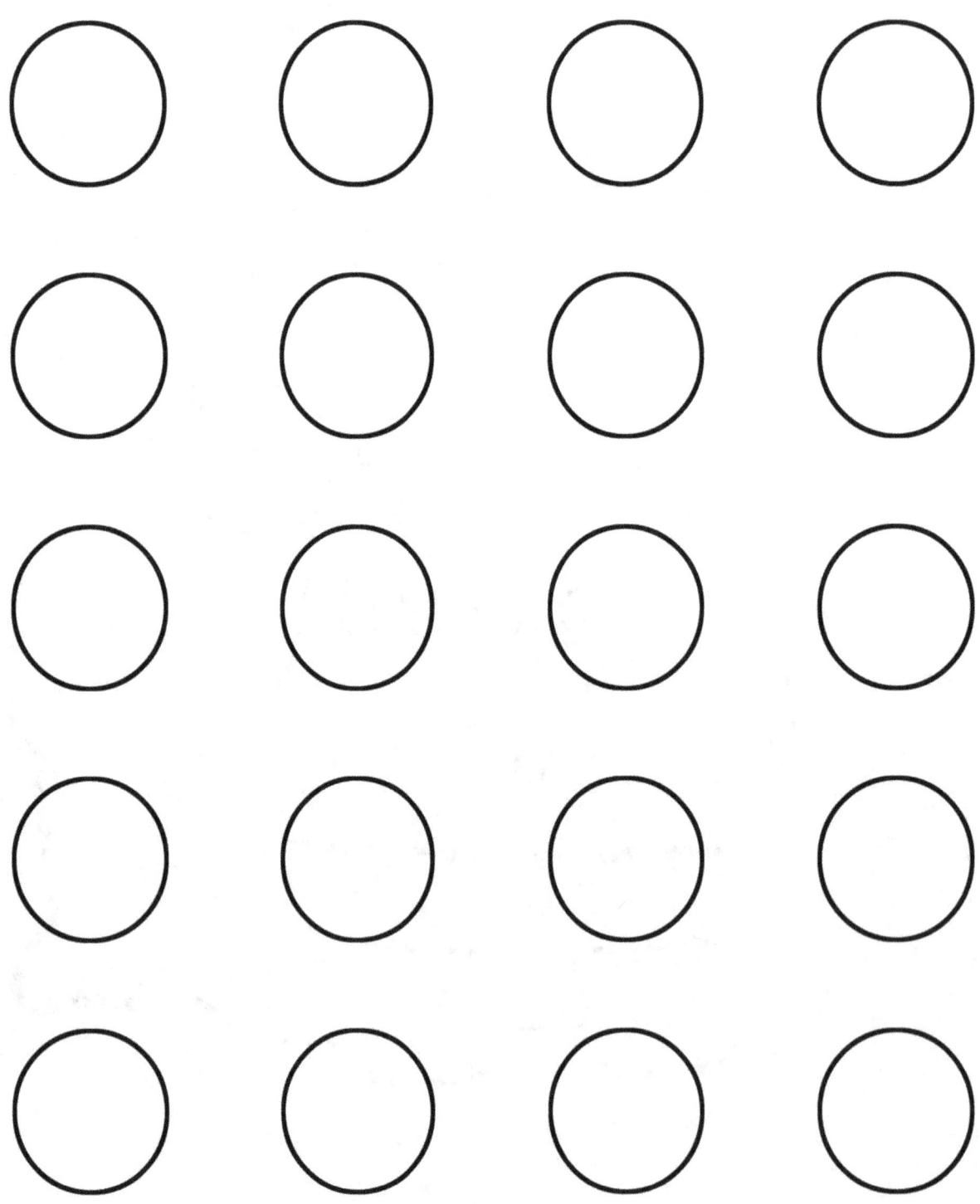

Continue to join the dots you have connected all the numbered dots. Then, color the picture!

Test Your Color

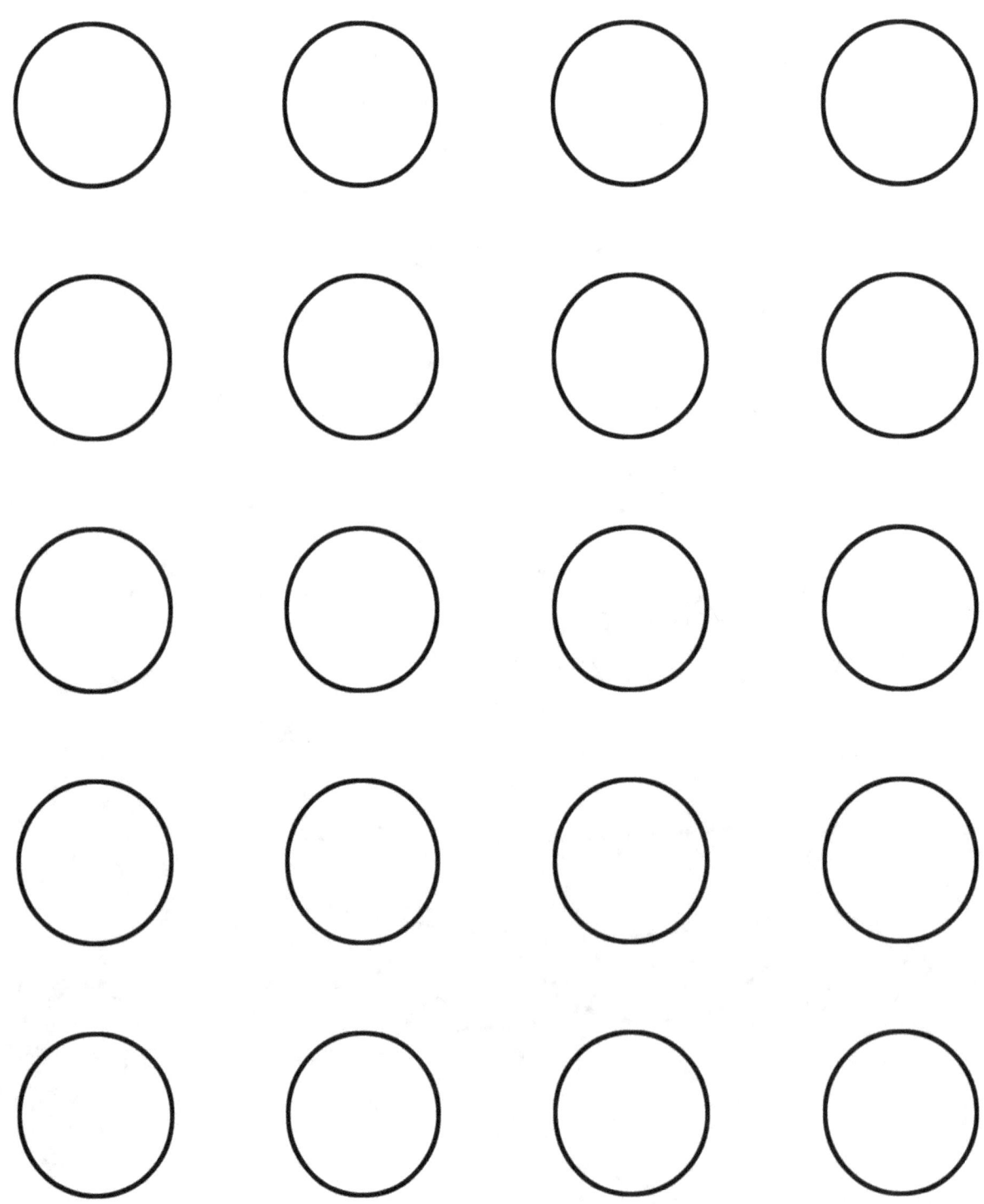

Continue to join the dots you have connected all the numbered dots. Then, color the picture!

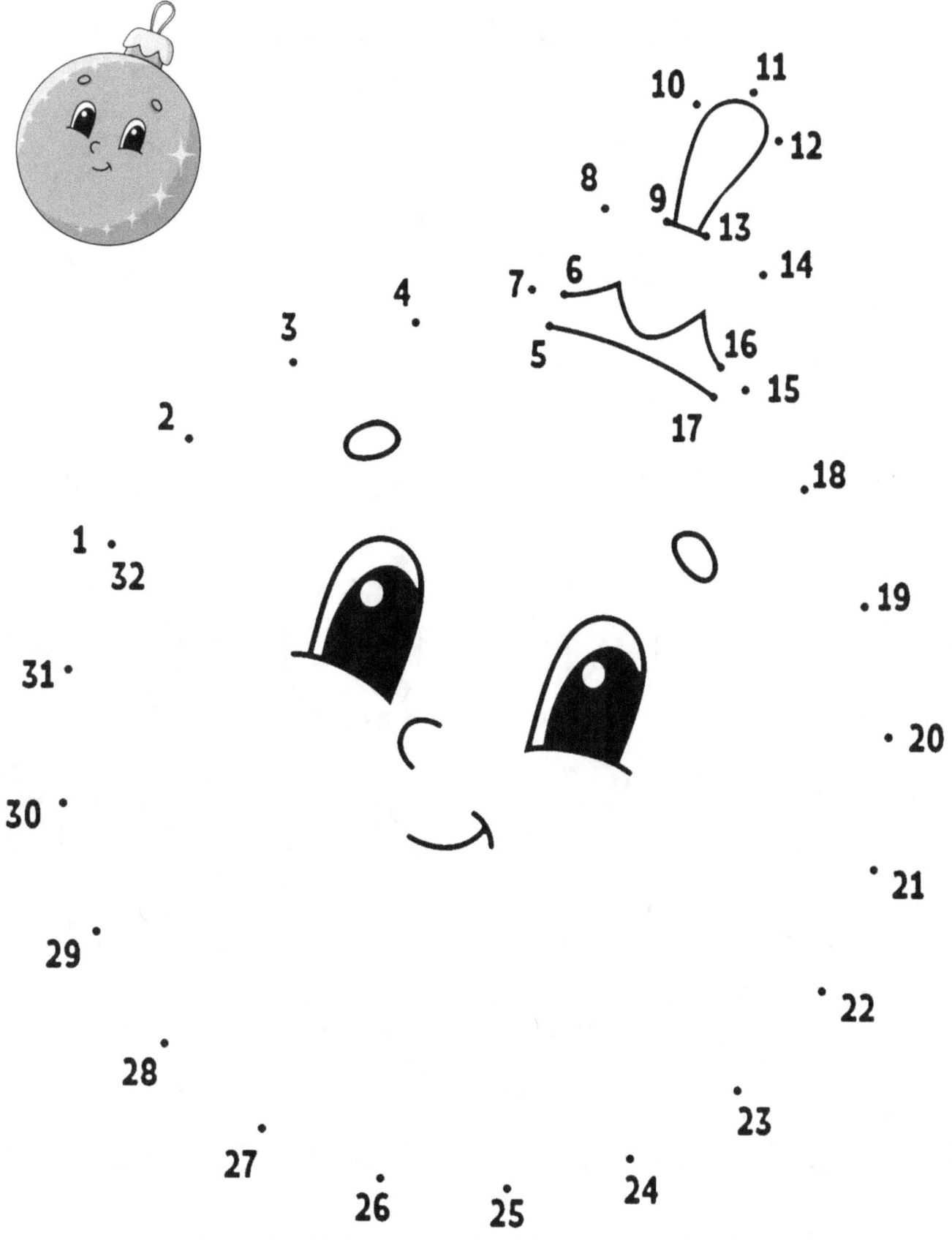

Test Your Color

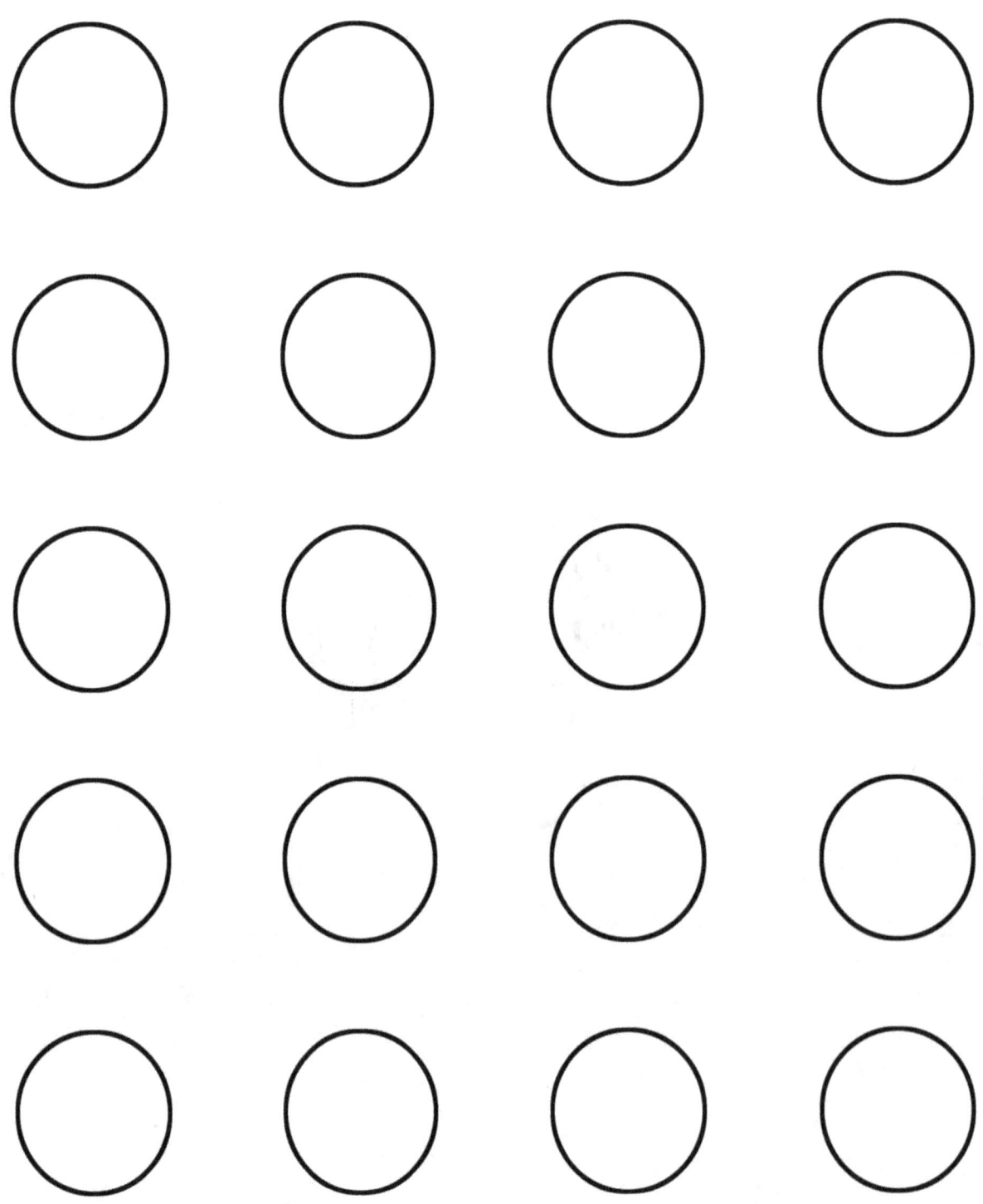

Test Your Color

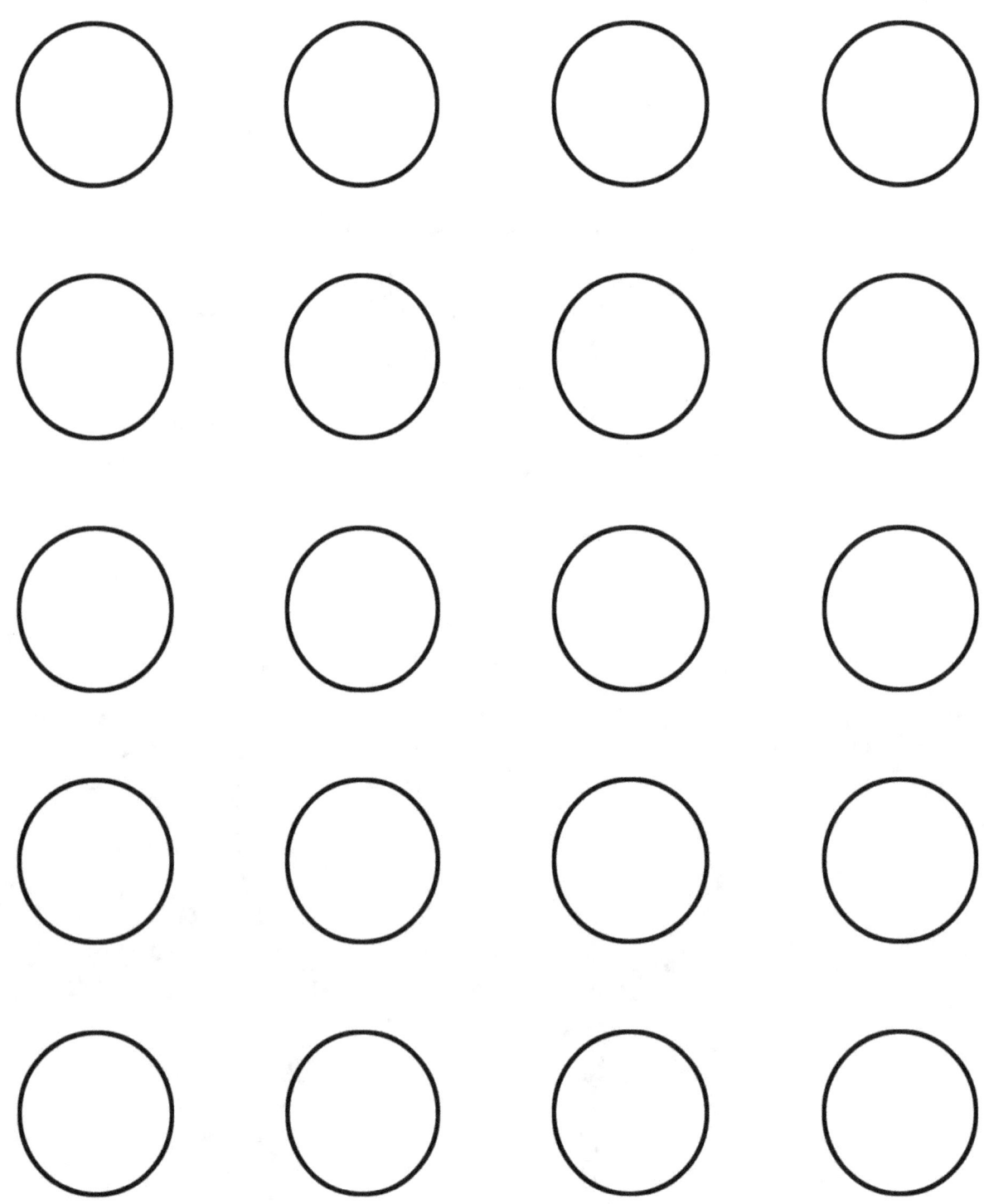

Continue to join the dots you have connected all the numbered dots. Then, color the picture!

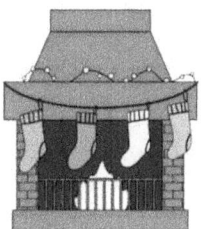

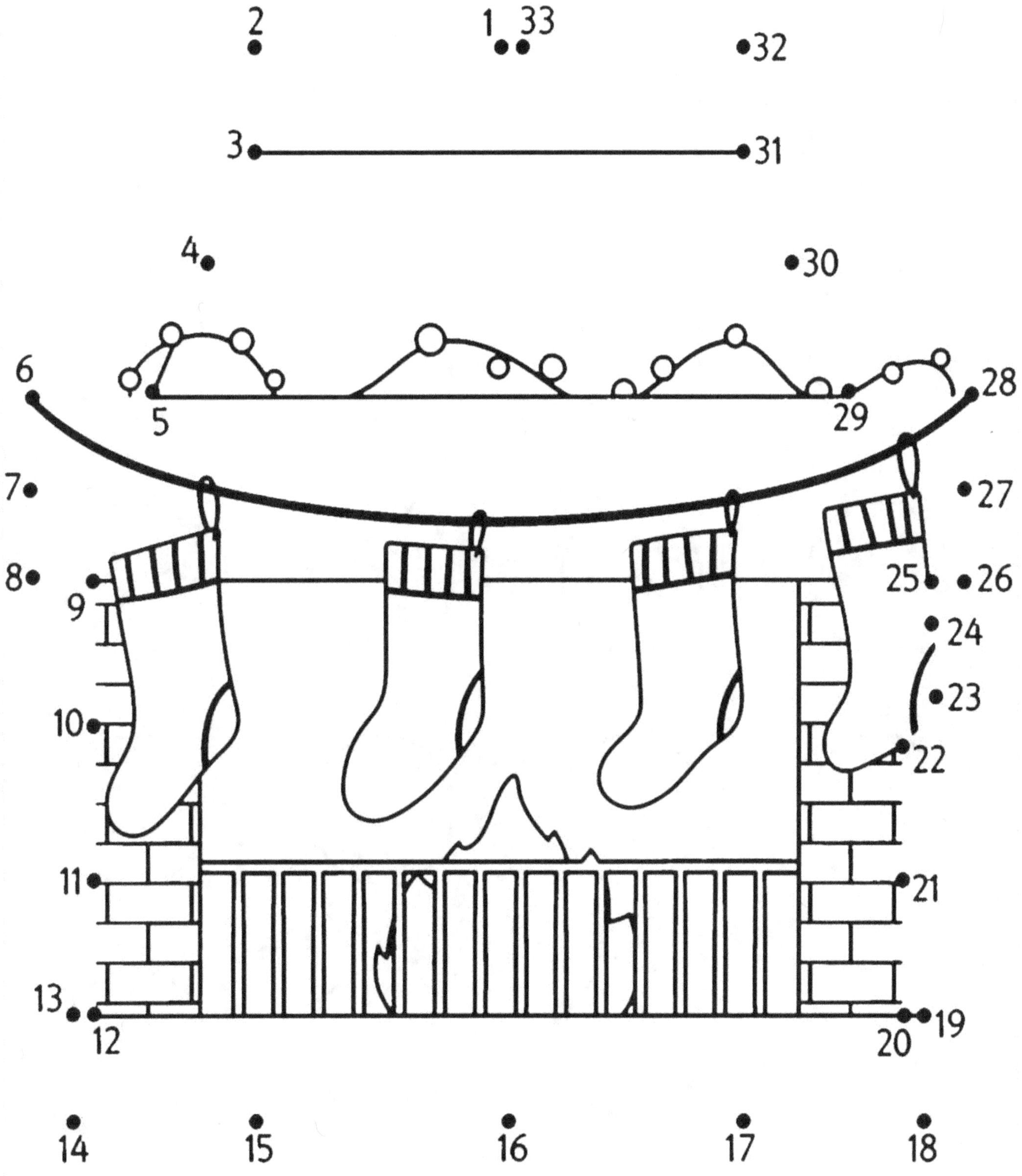

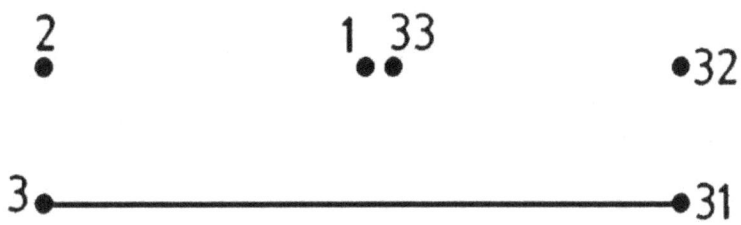

Test Your Color

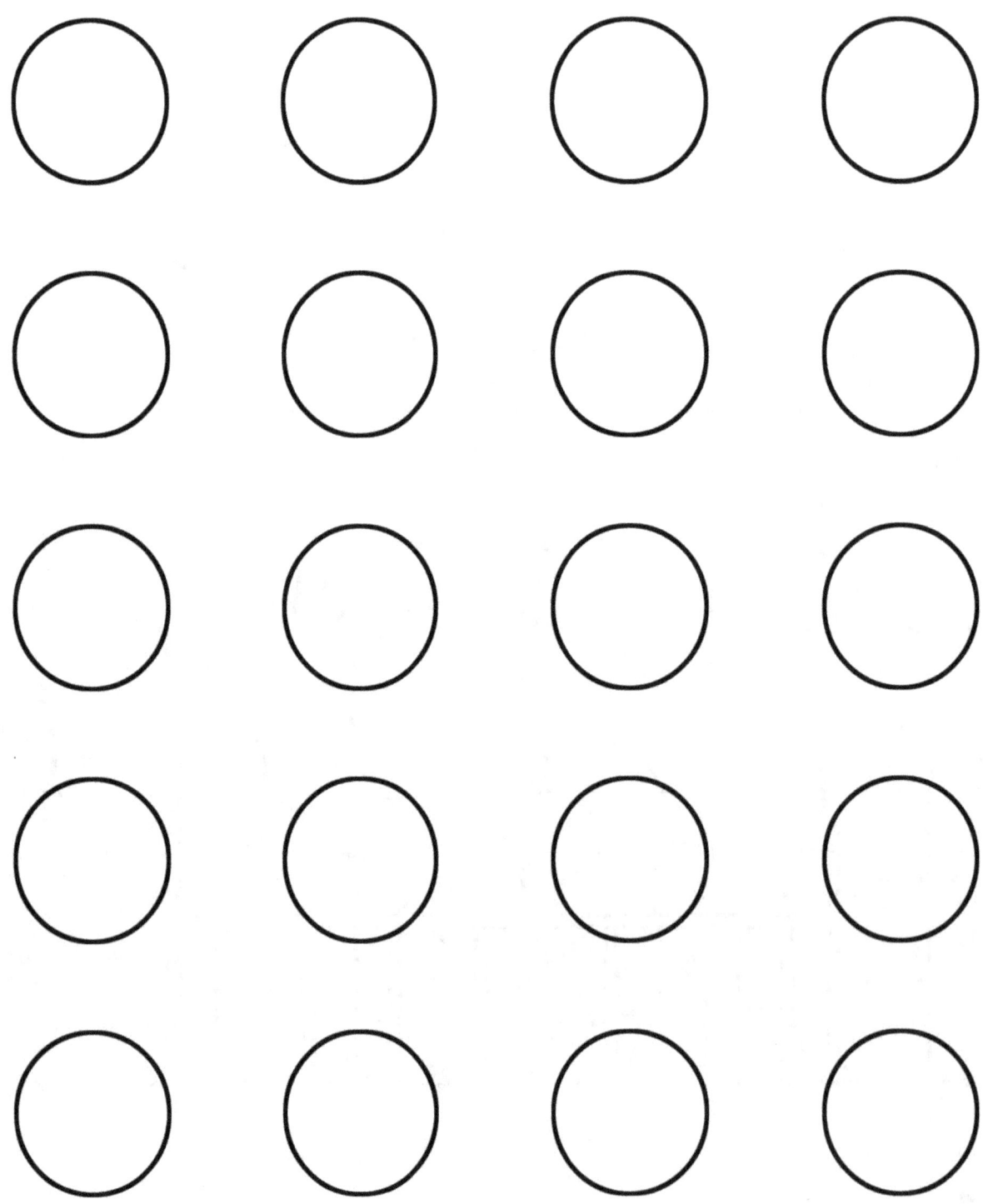

Continue to join the dots you have connected all the numbered dots. Then, color the picture!

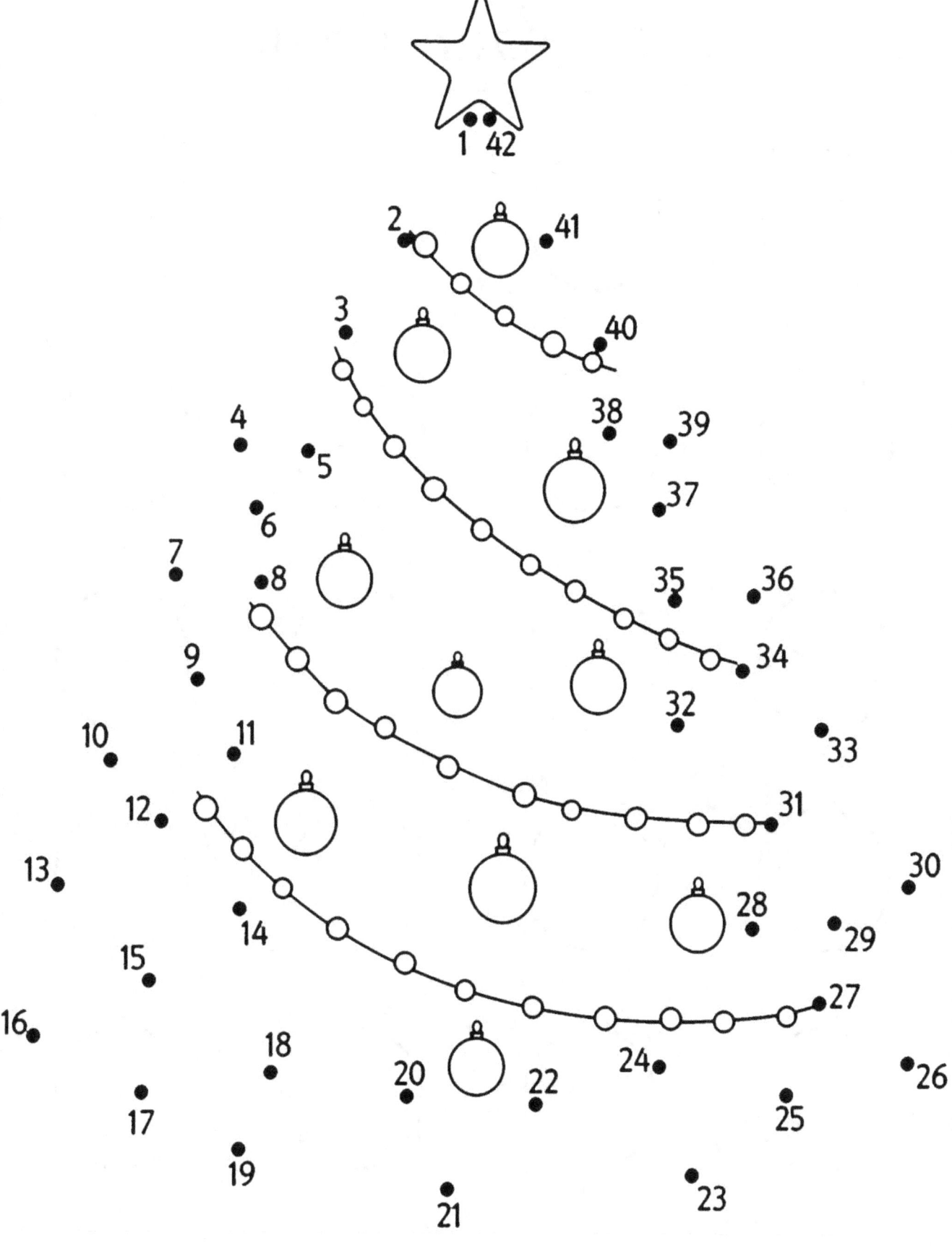

Test Your Color

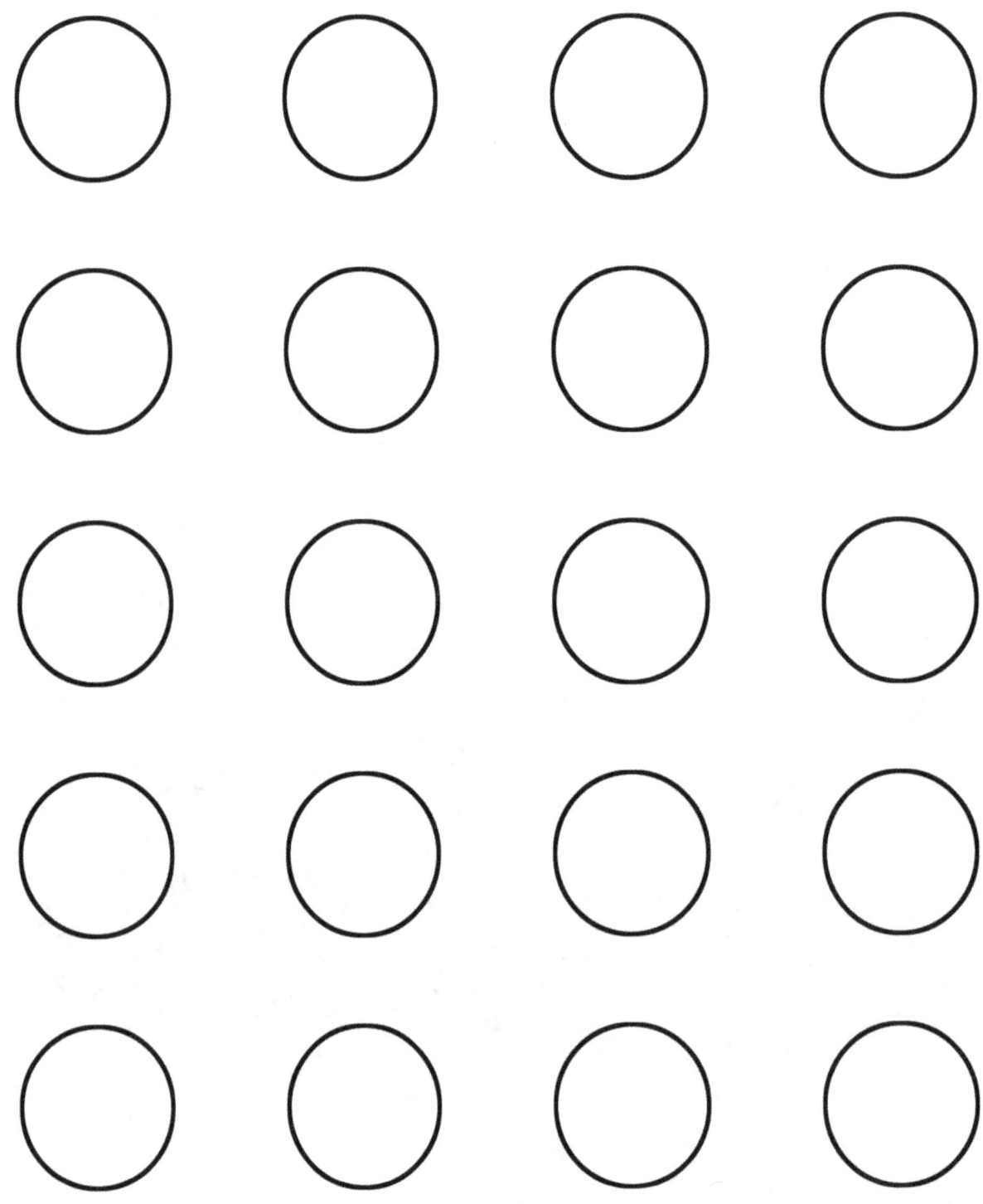

Continue to join the dots you have connected all the numbered dots. Then, color the picture!

Test Your Color

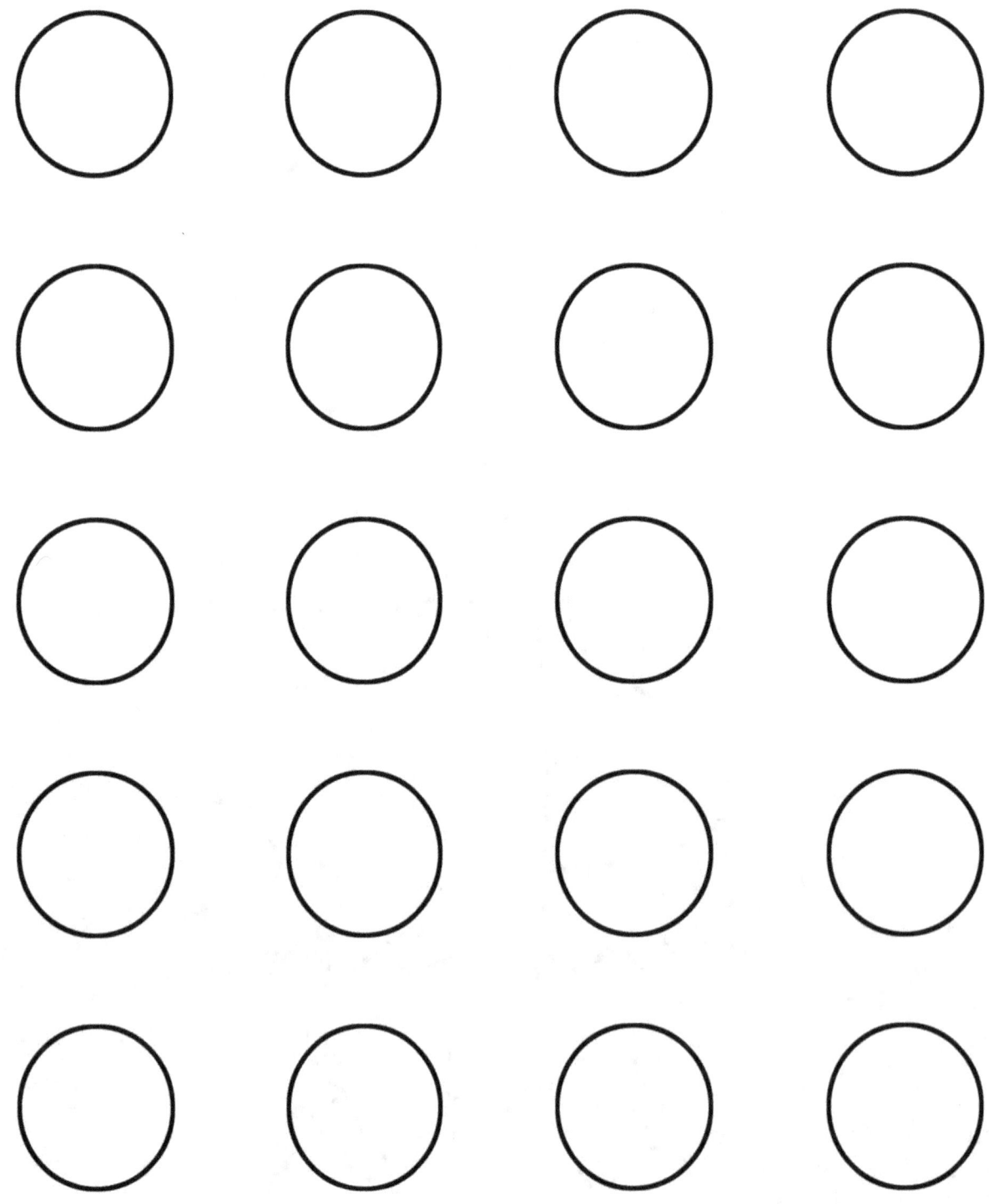

Continue to join the dots you have connected all the numbered dots. Then, color the picture!

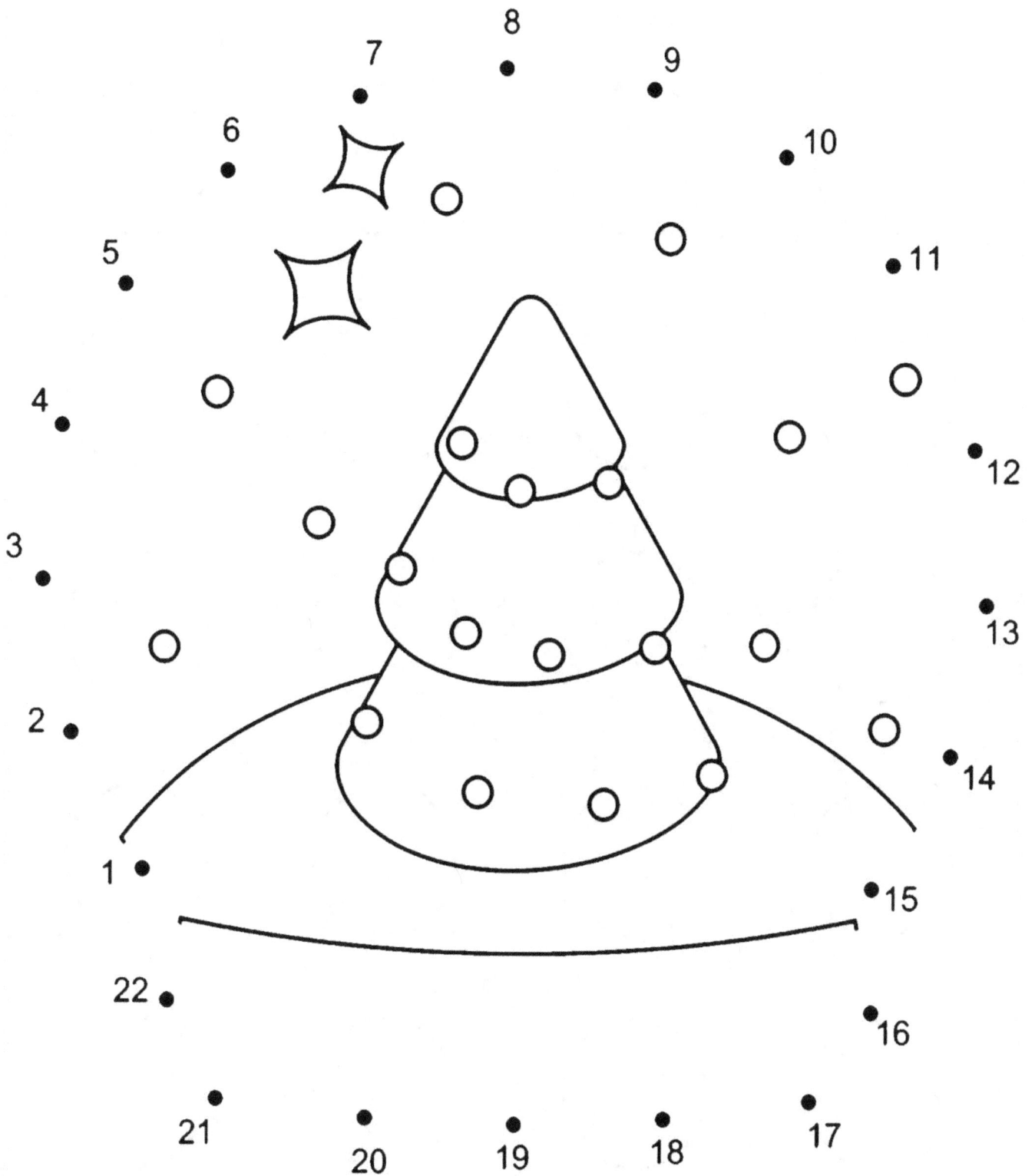

Test Your Color

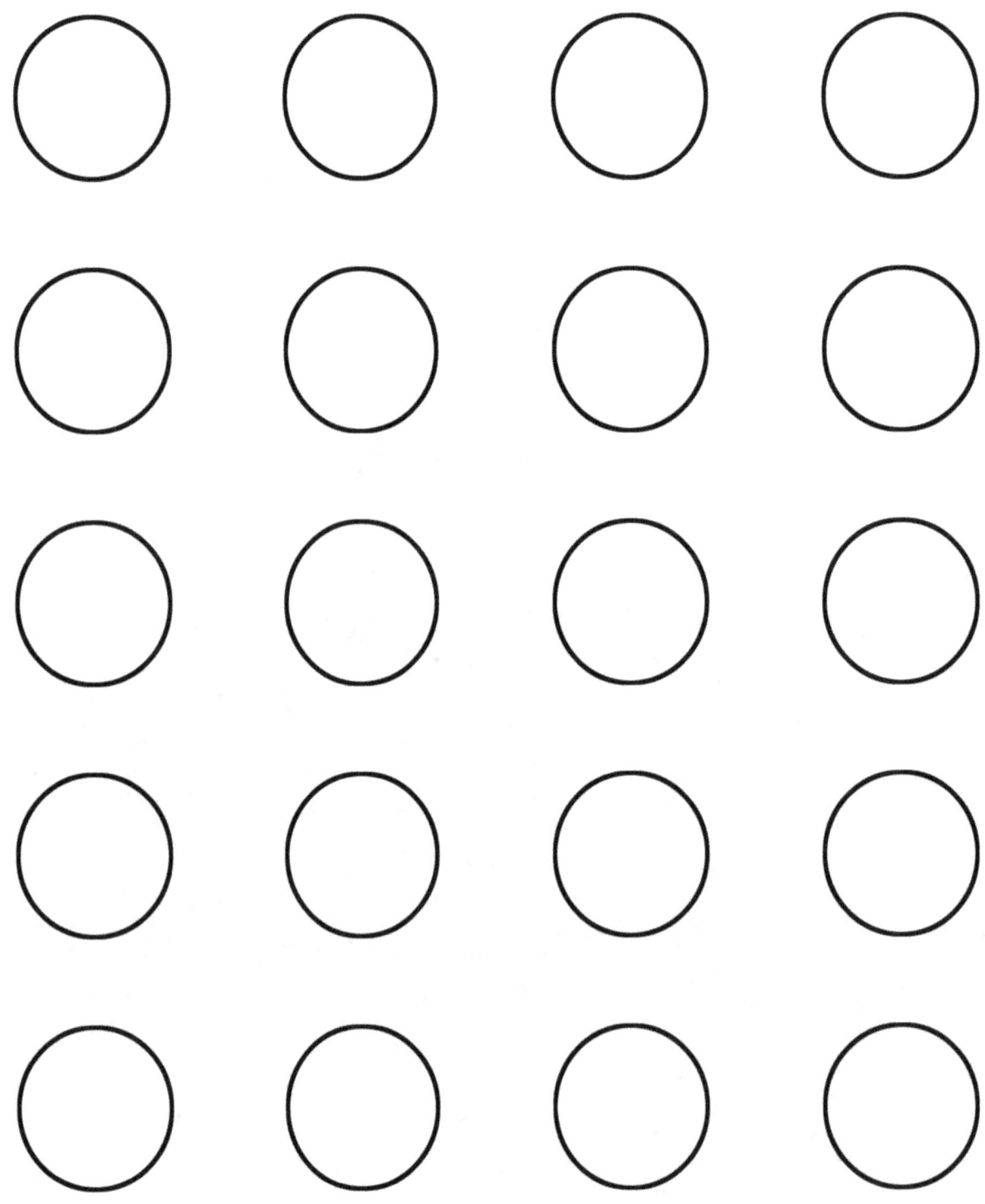

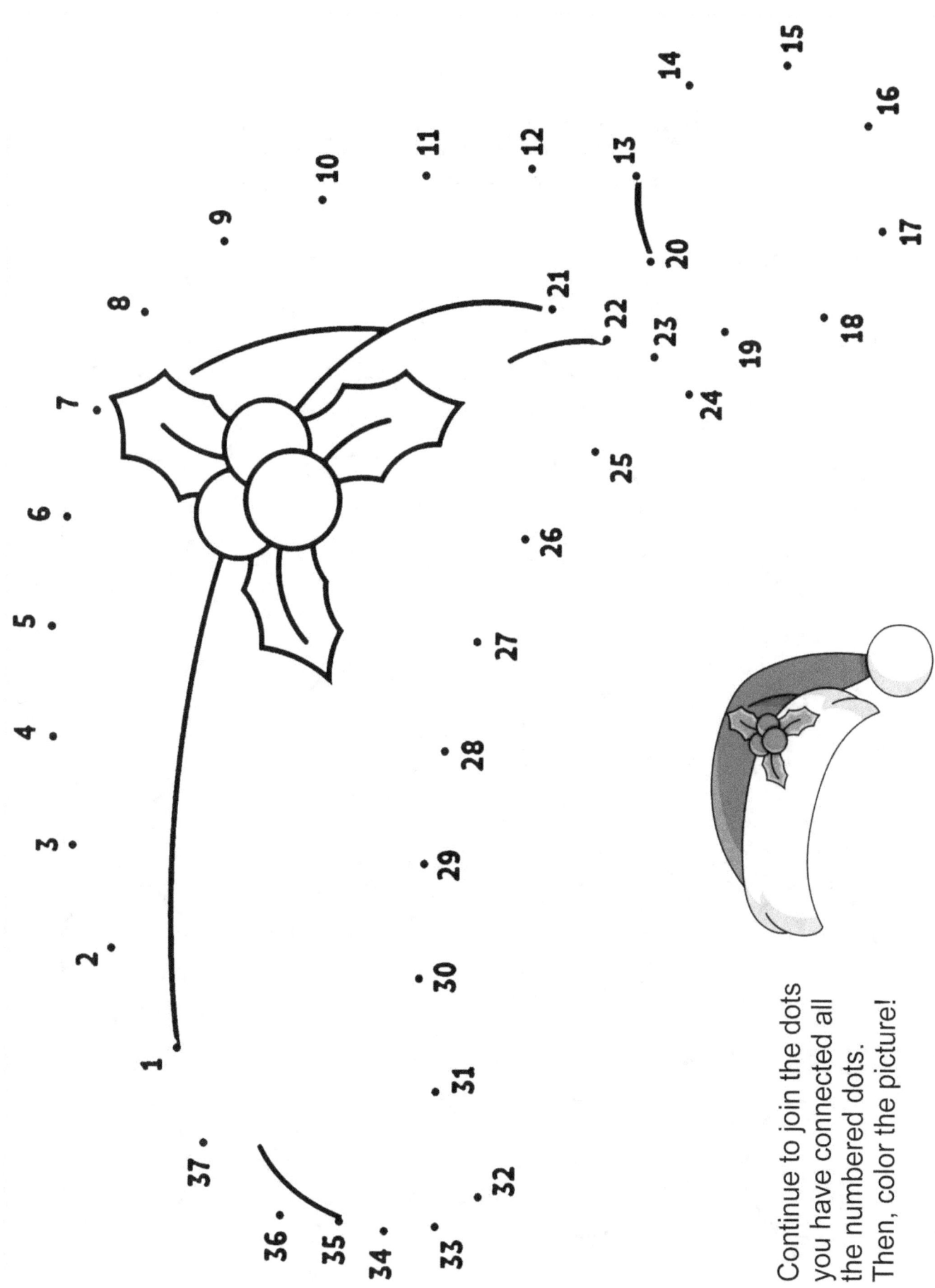

Test Your Color

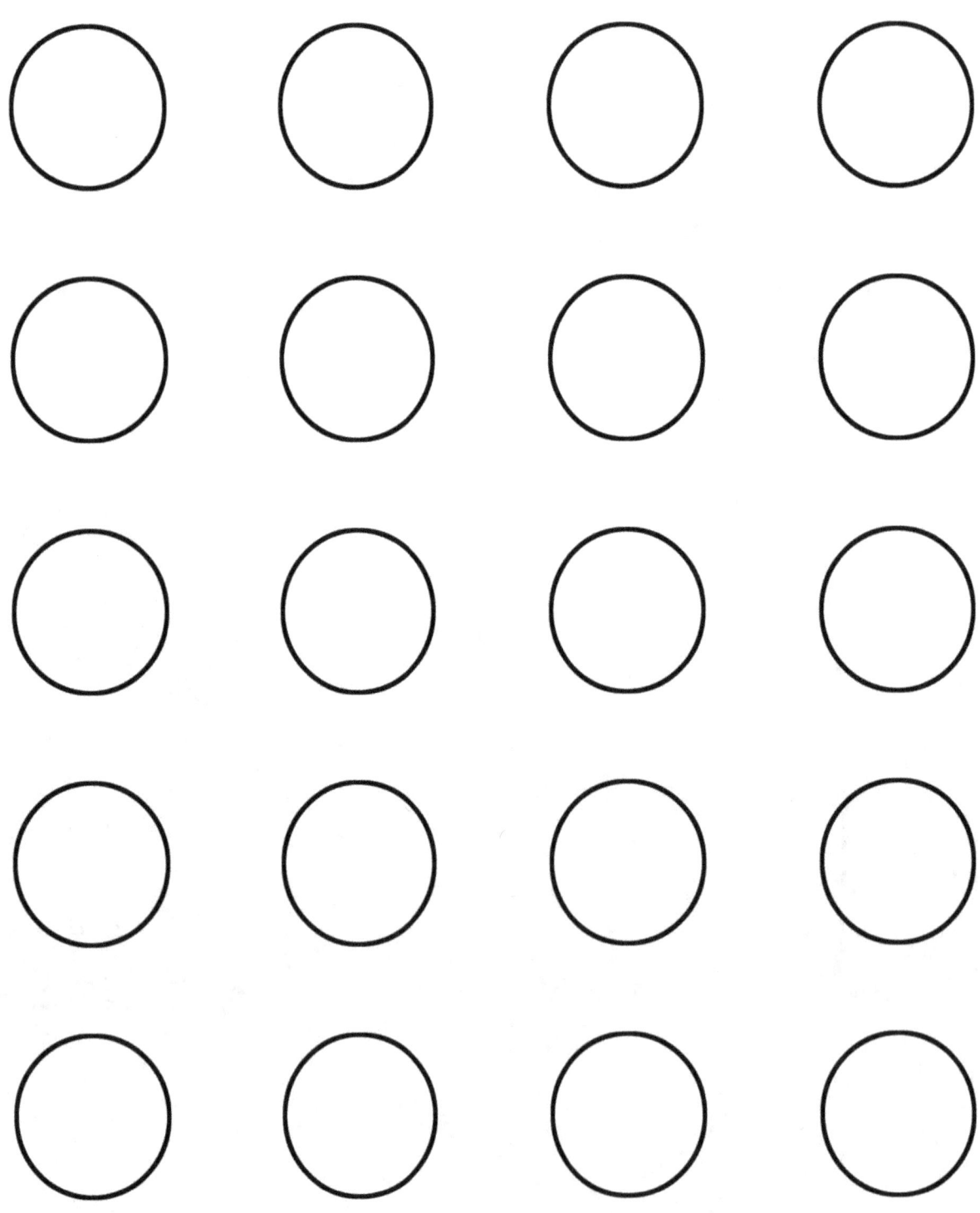

Continue to join the dots you have connected all the numbered dots. Then, color the picture!

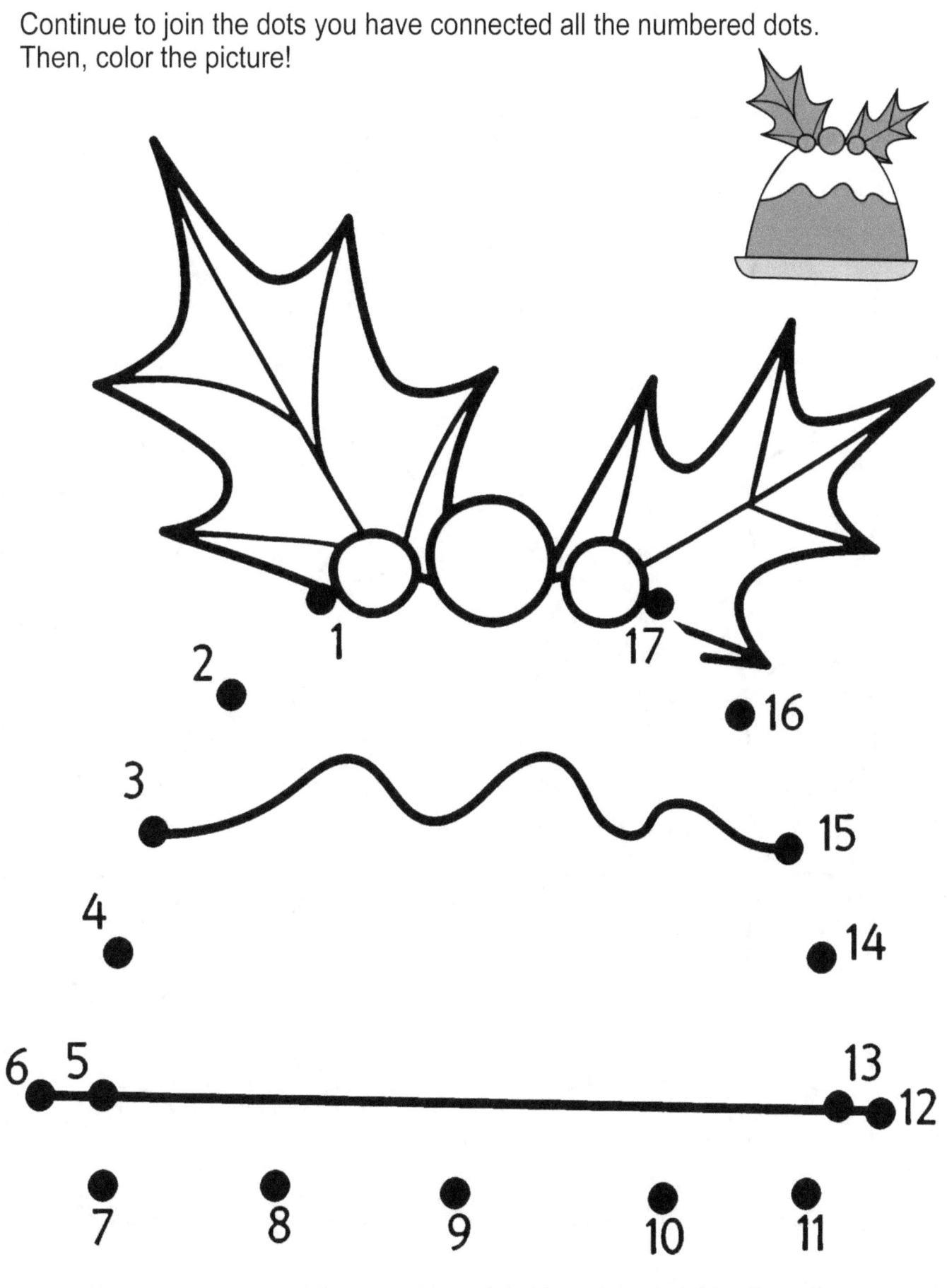

Test Your Color

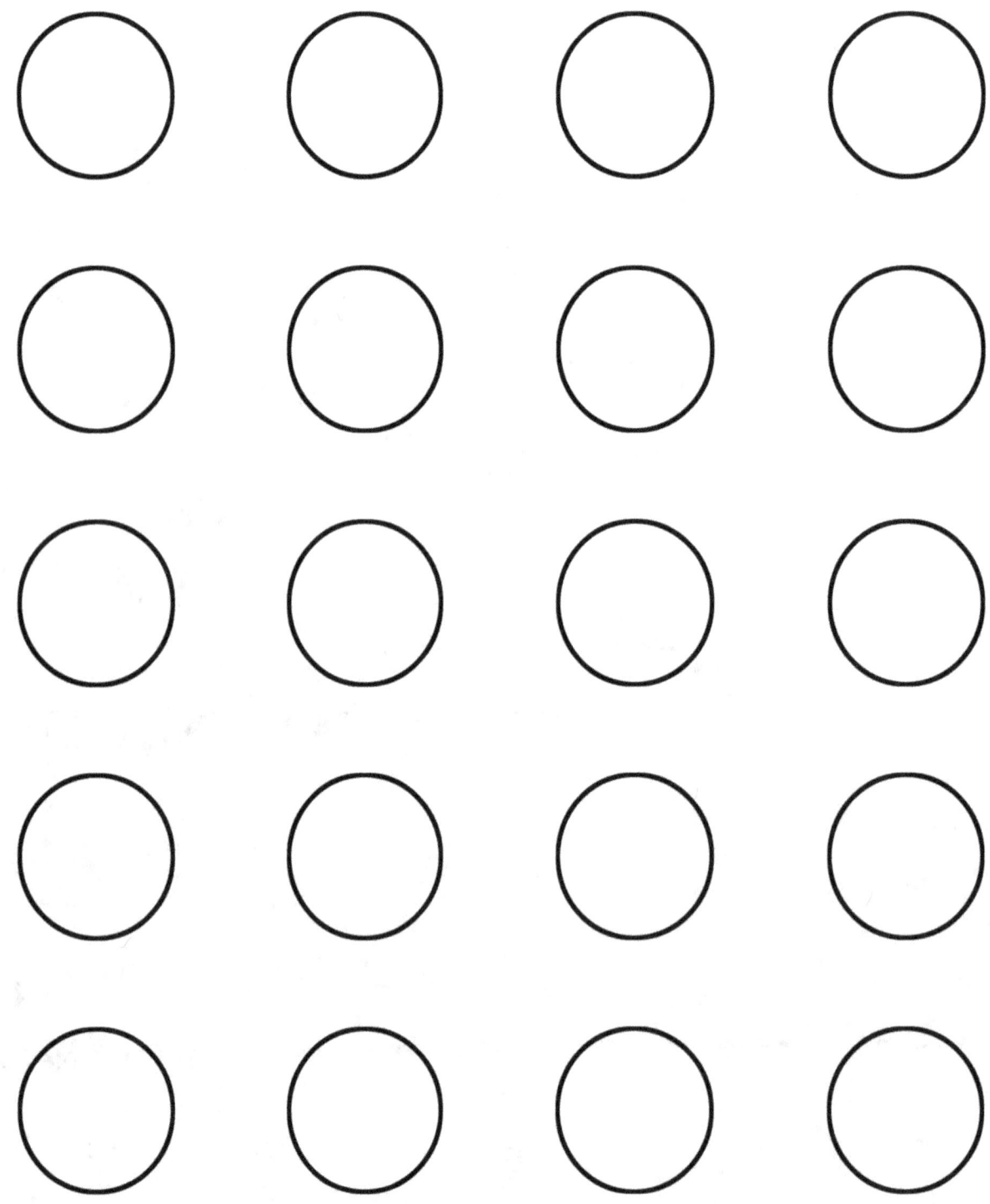

Continue to join the dots you have connected all the numbered dots. Then, color the picture!

Test Your Color

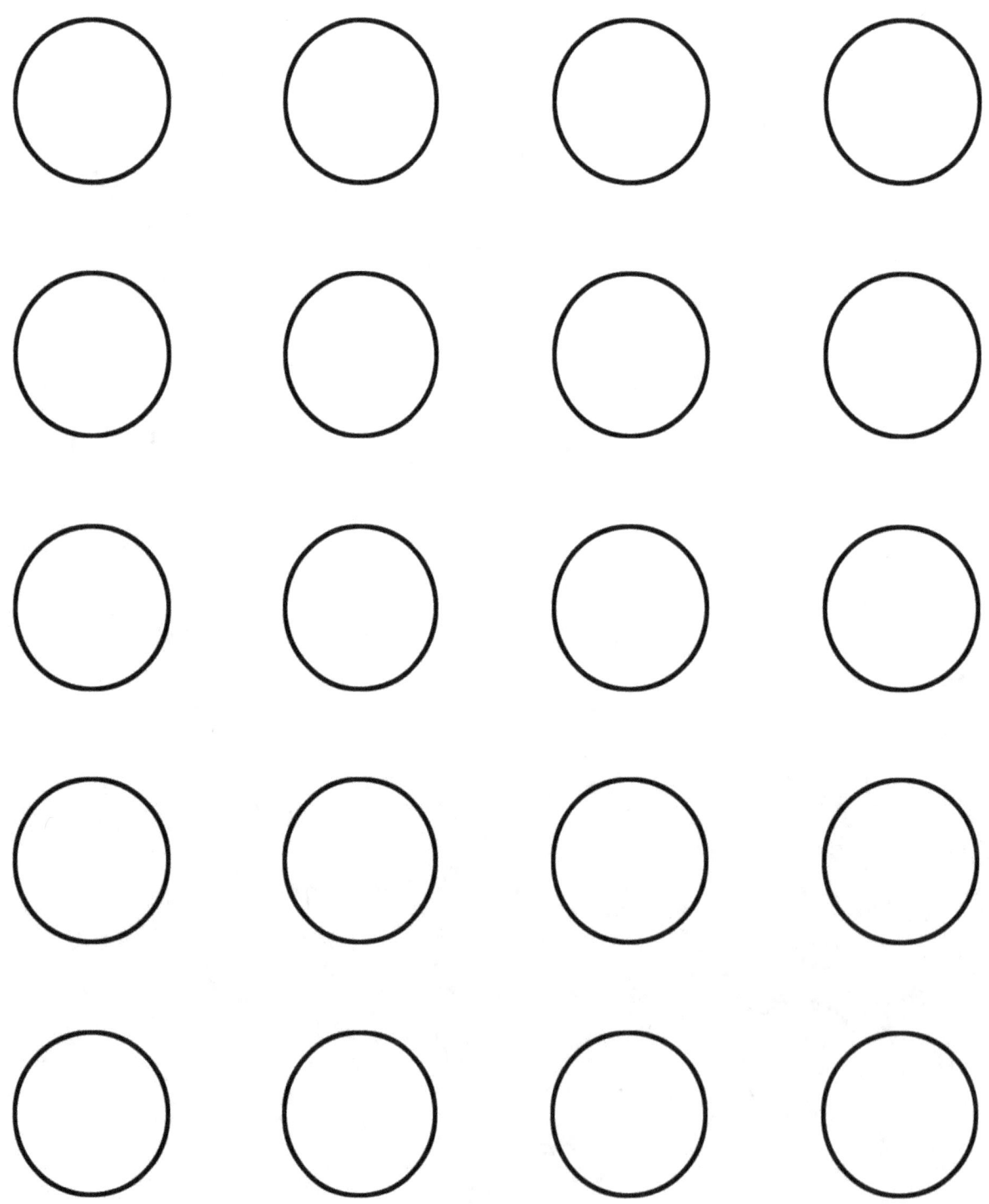

Test Your Color

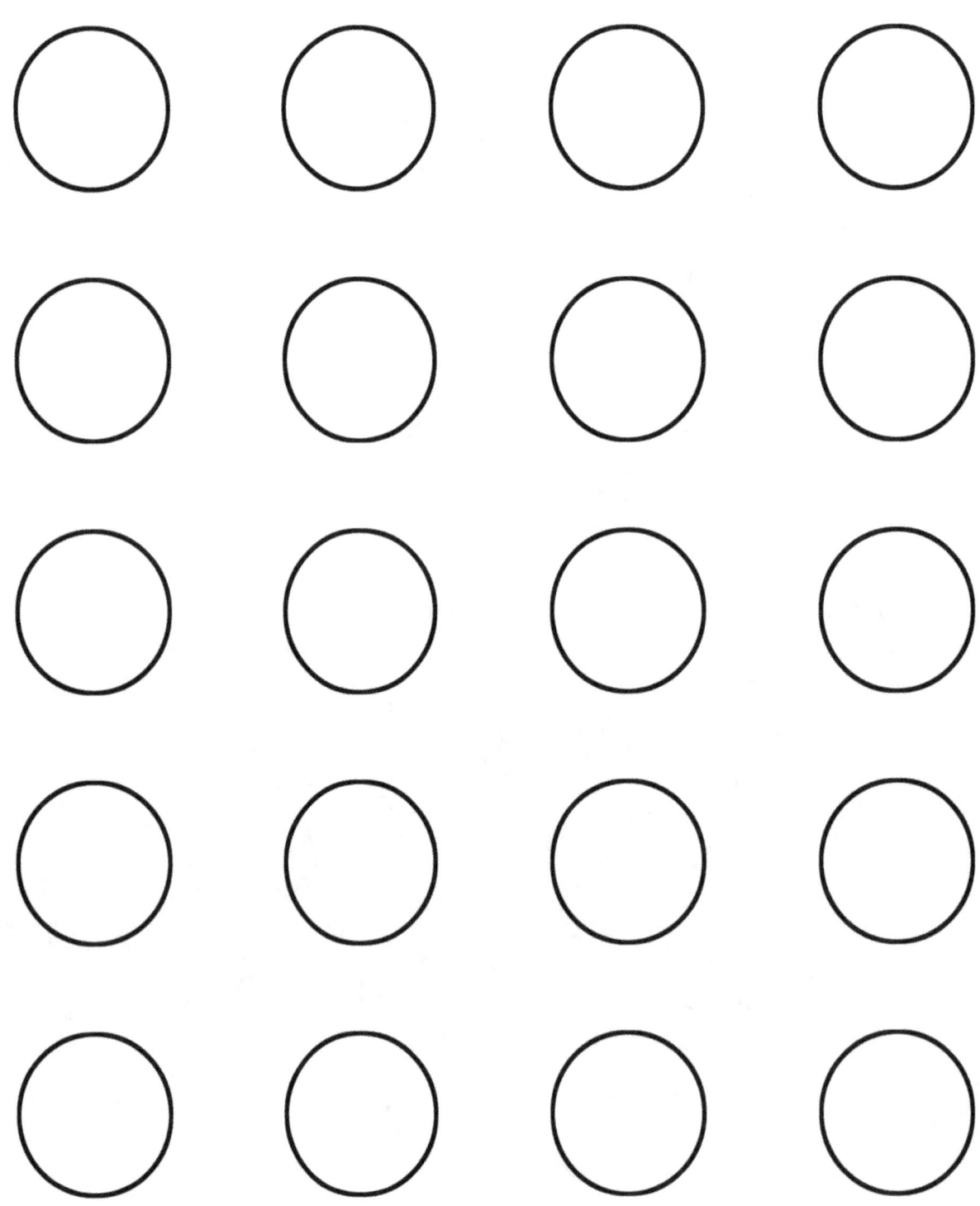

Continue to join the dots you have connected all the numbered dots. Then, color the picture!

Test Your Color

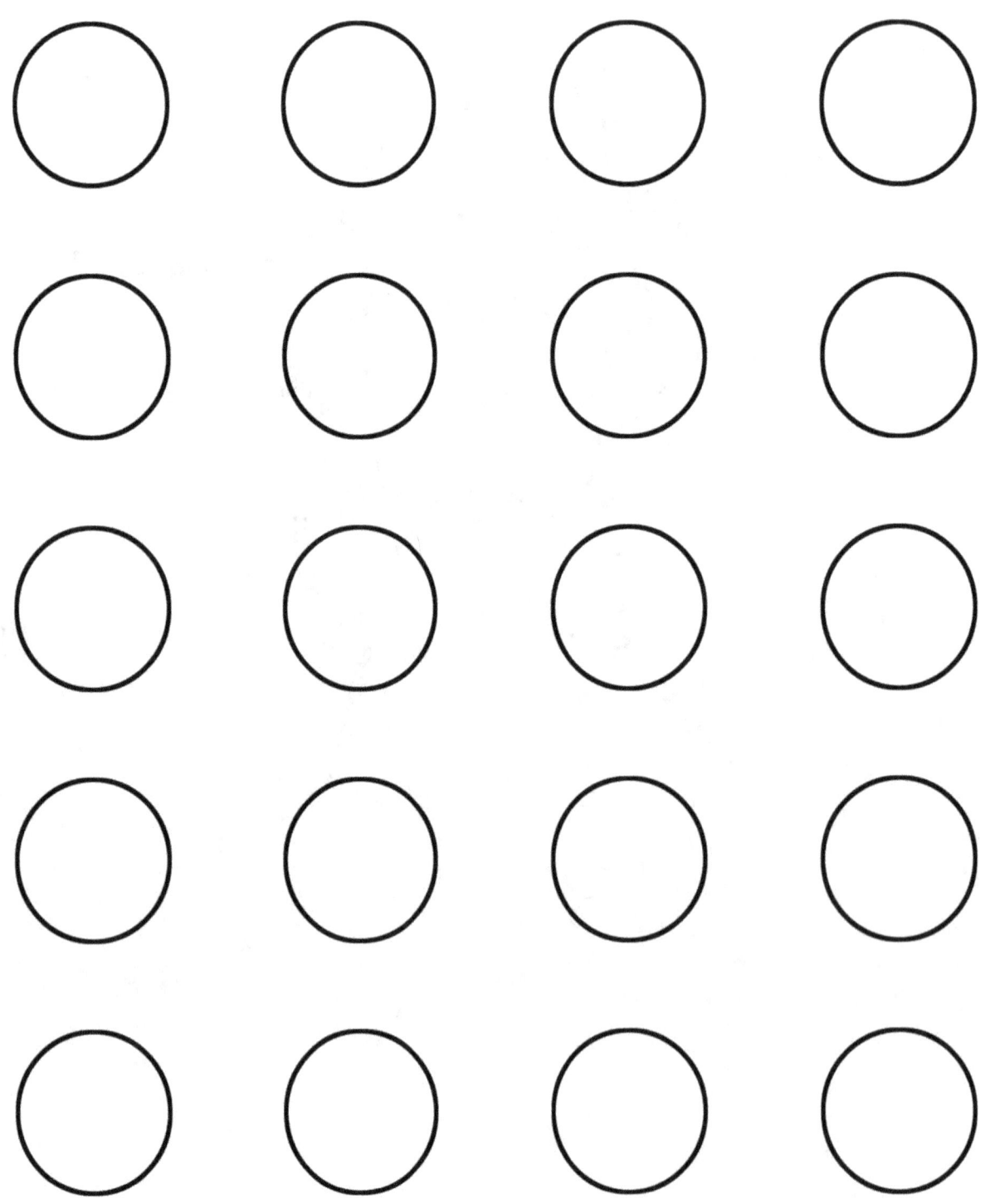

Continue to join the dots you have connected all the numbered dots. Then, color the picture!

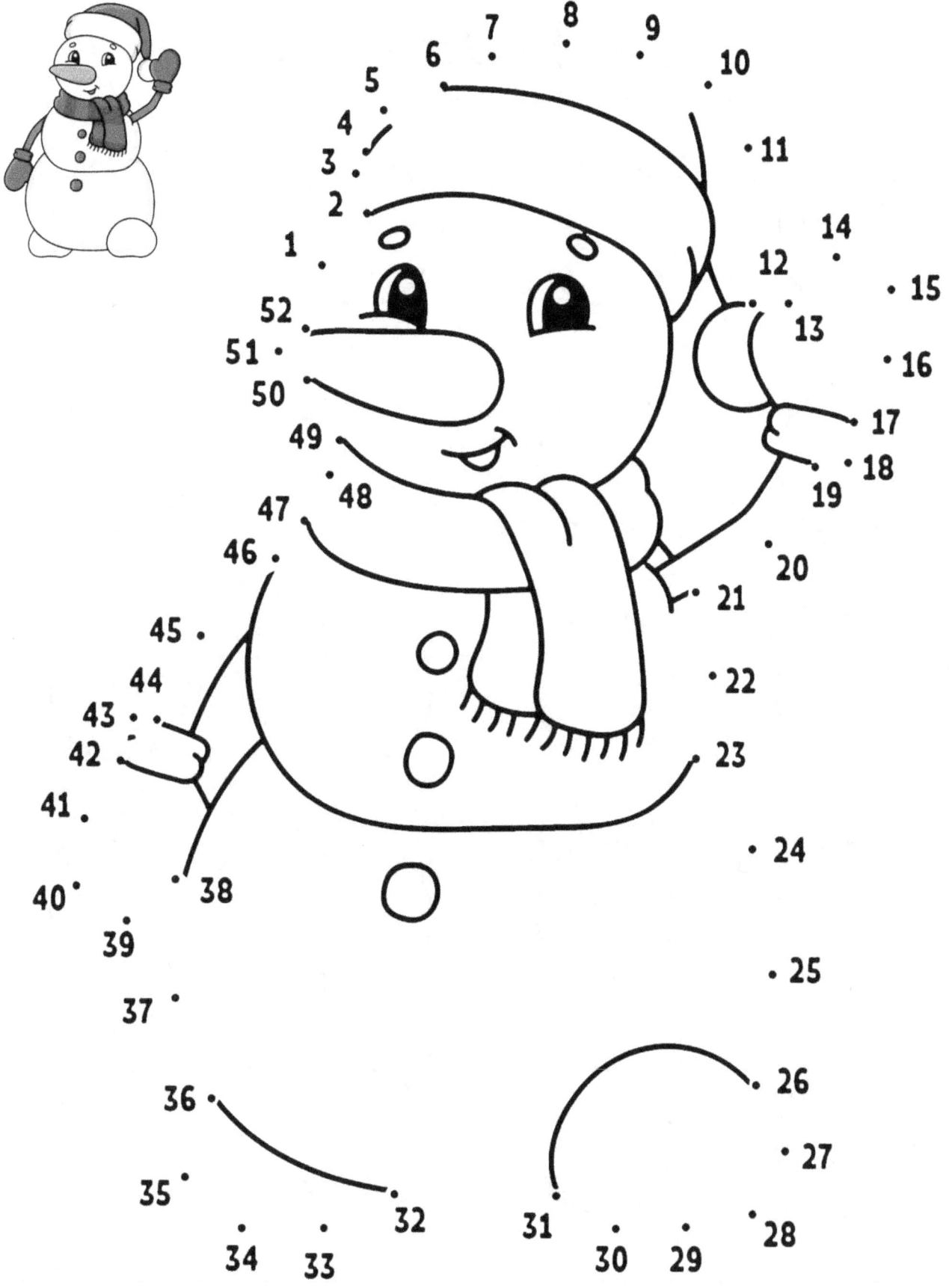

Test Your Color

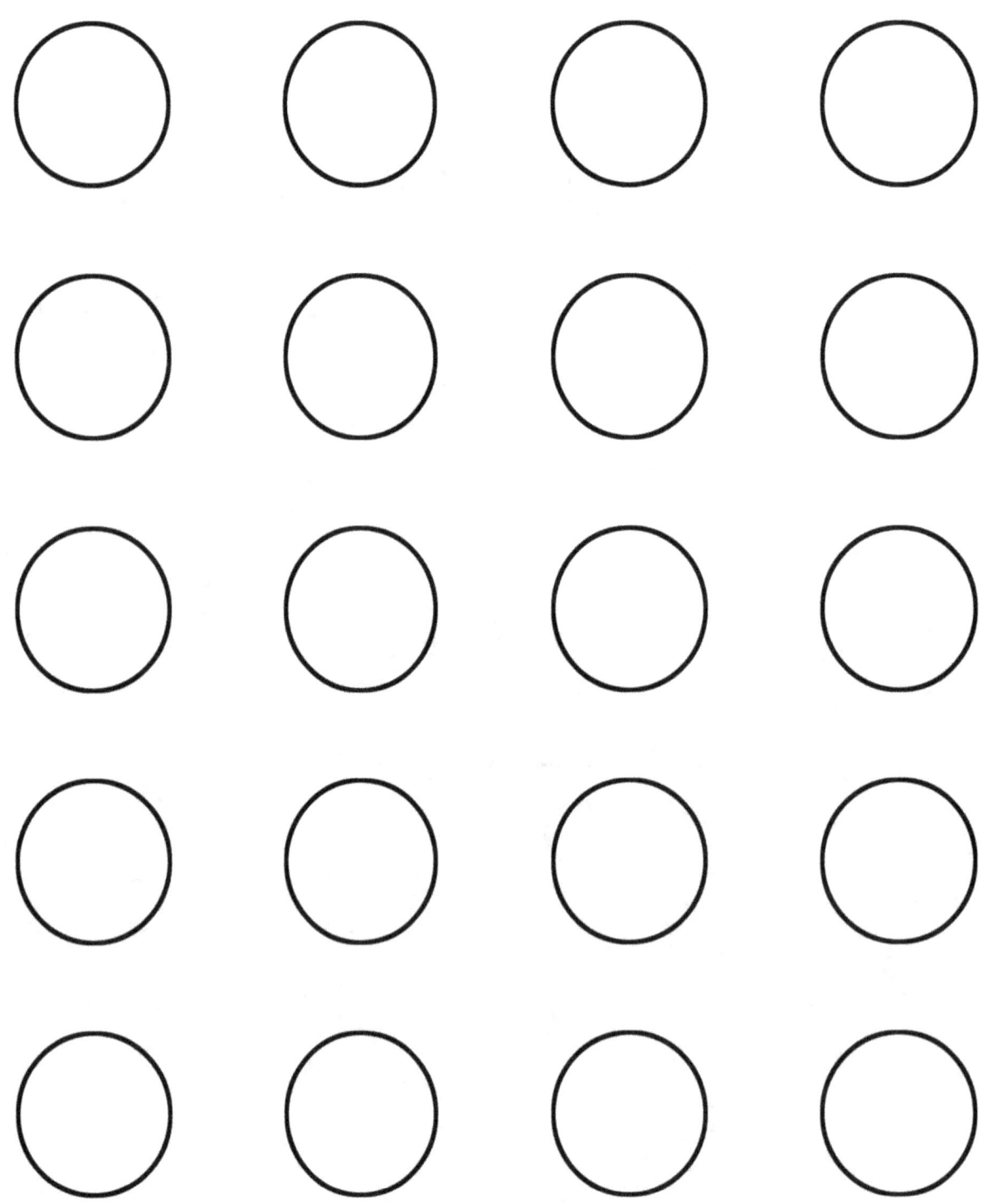

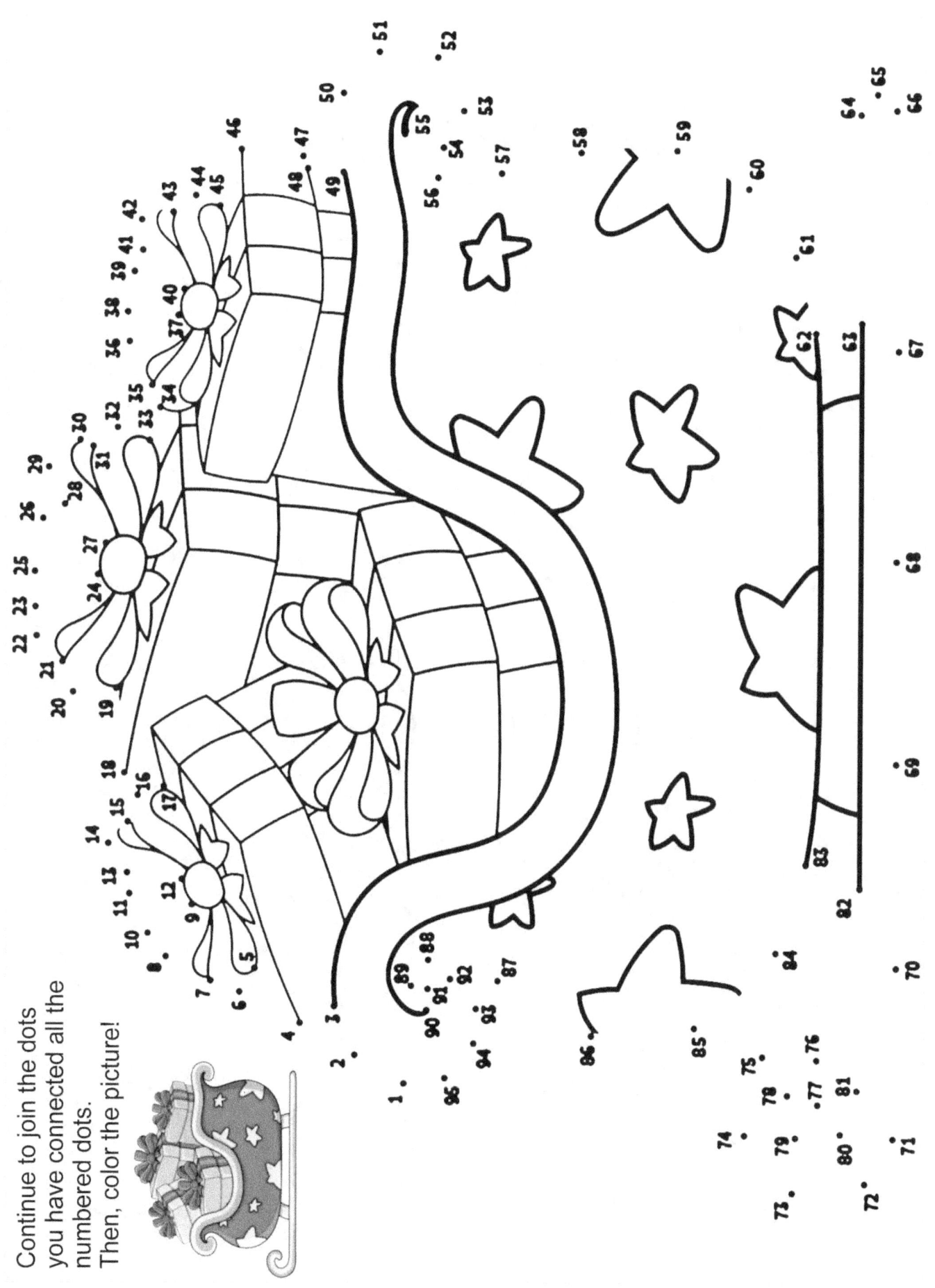

Test Your Color

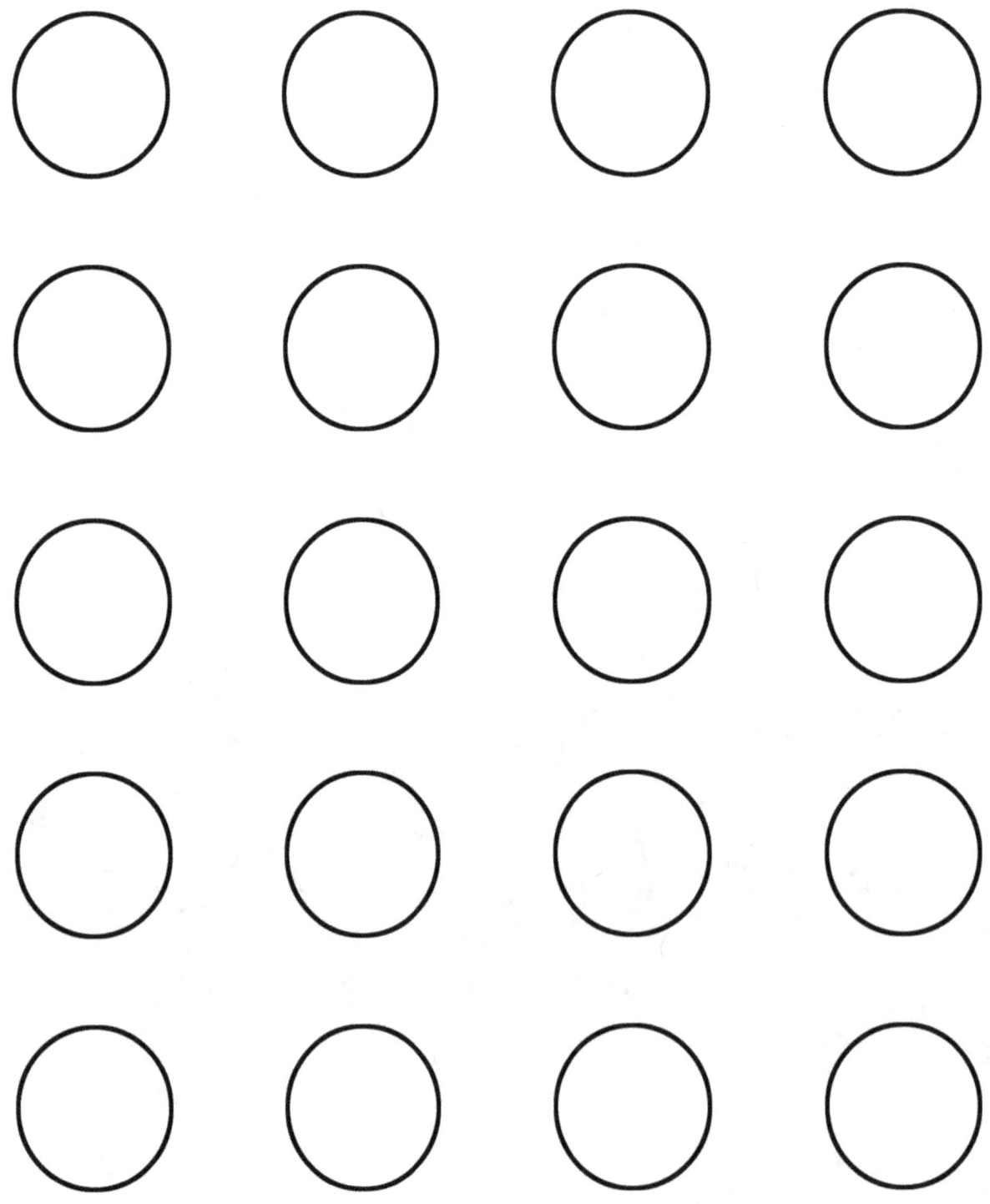

Continue to join the dots you have connected all the numbered dots. Then, color the picture!

Test Your Color

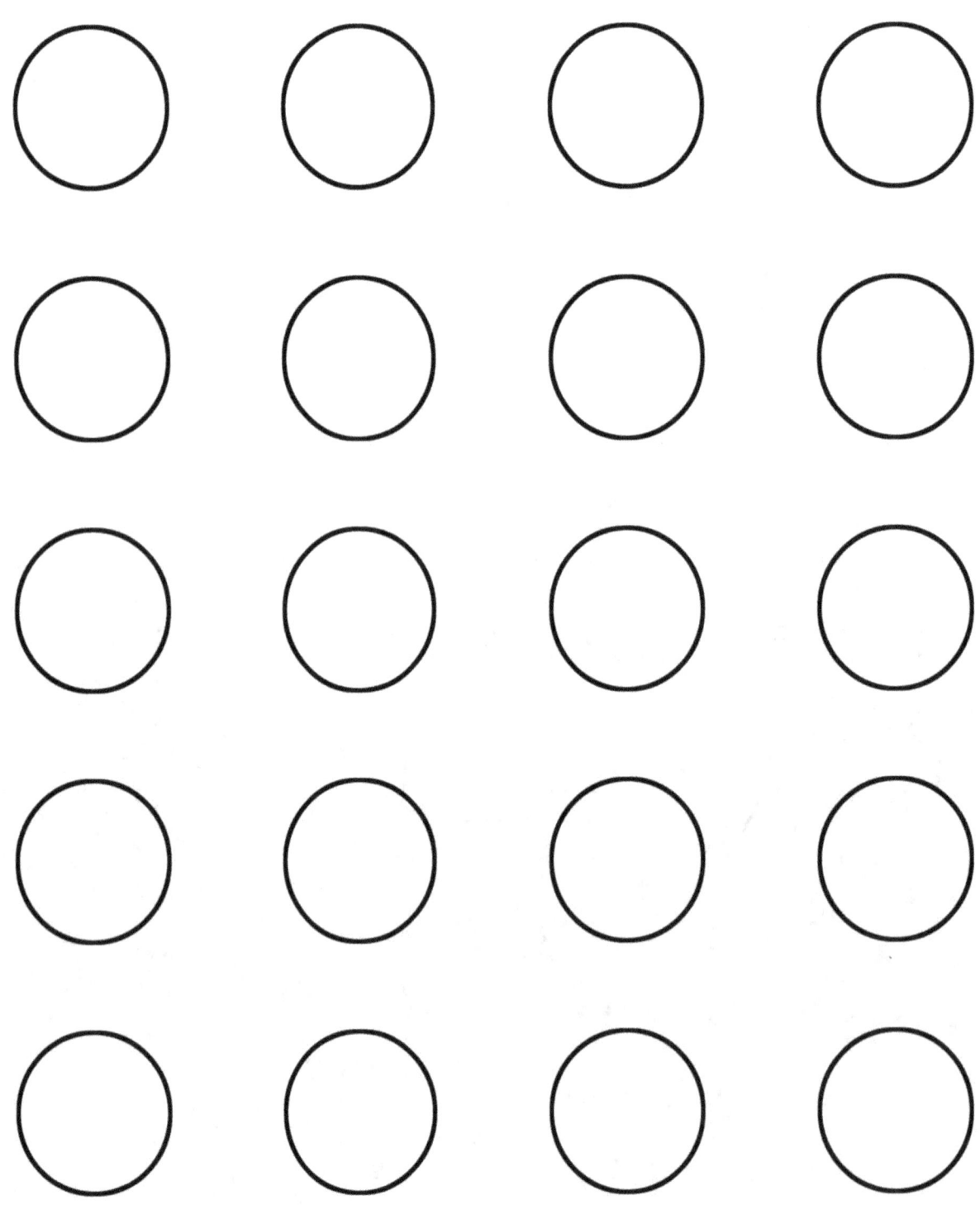

Continue to join the dots you have connected all the numbered dots. Then, color the picture!

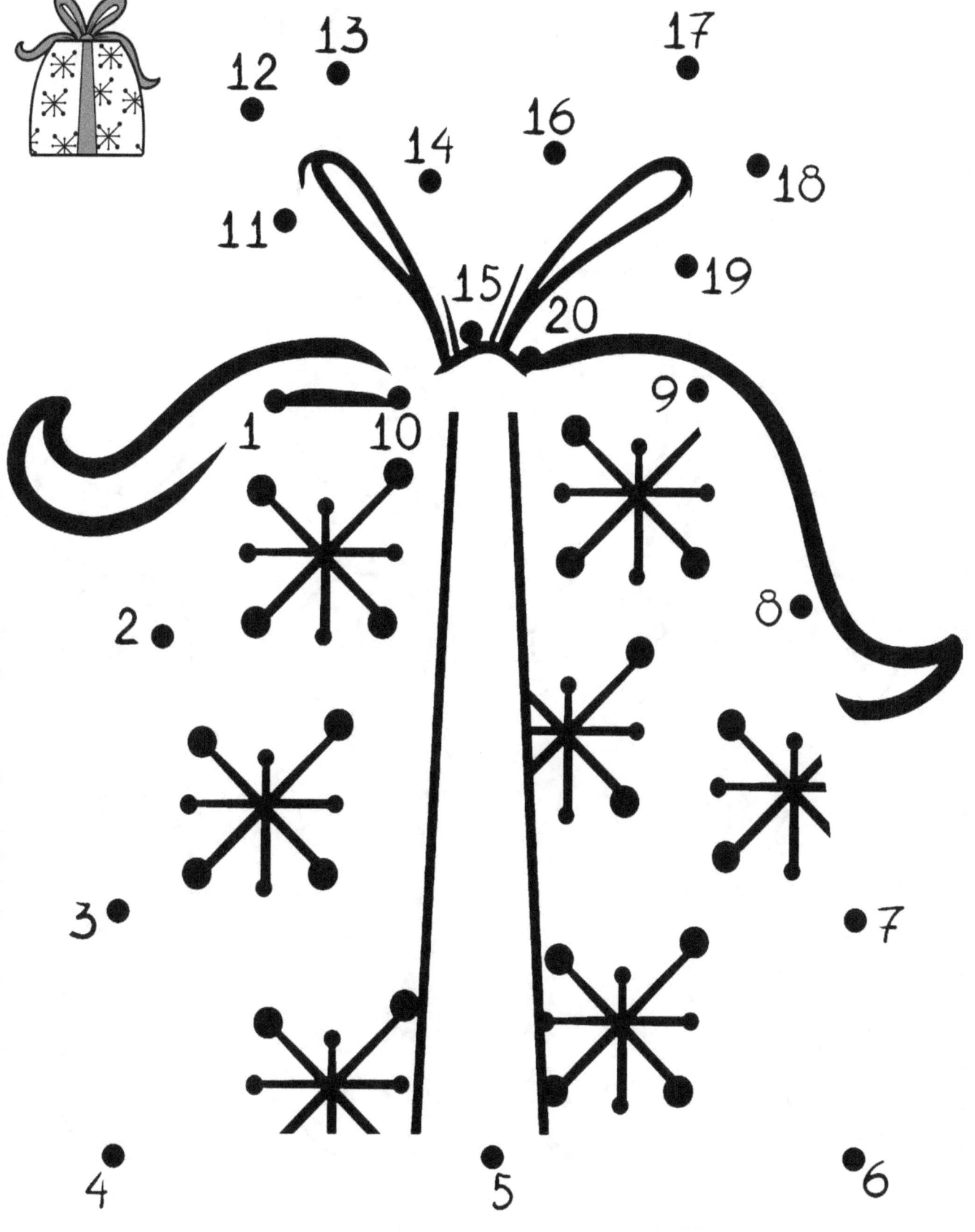

Test Your Color

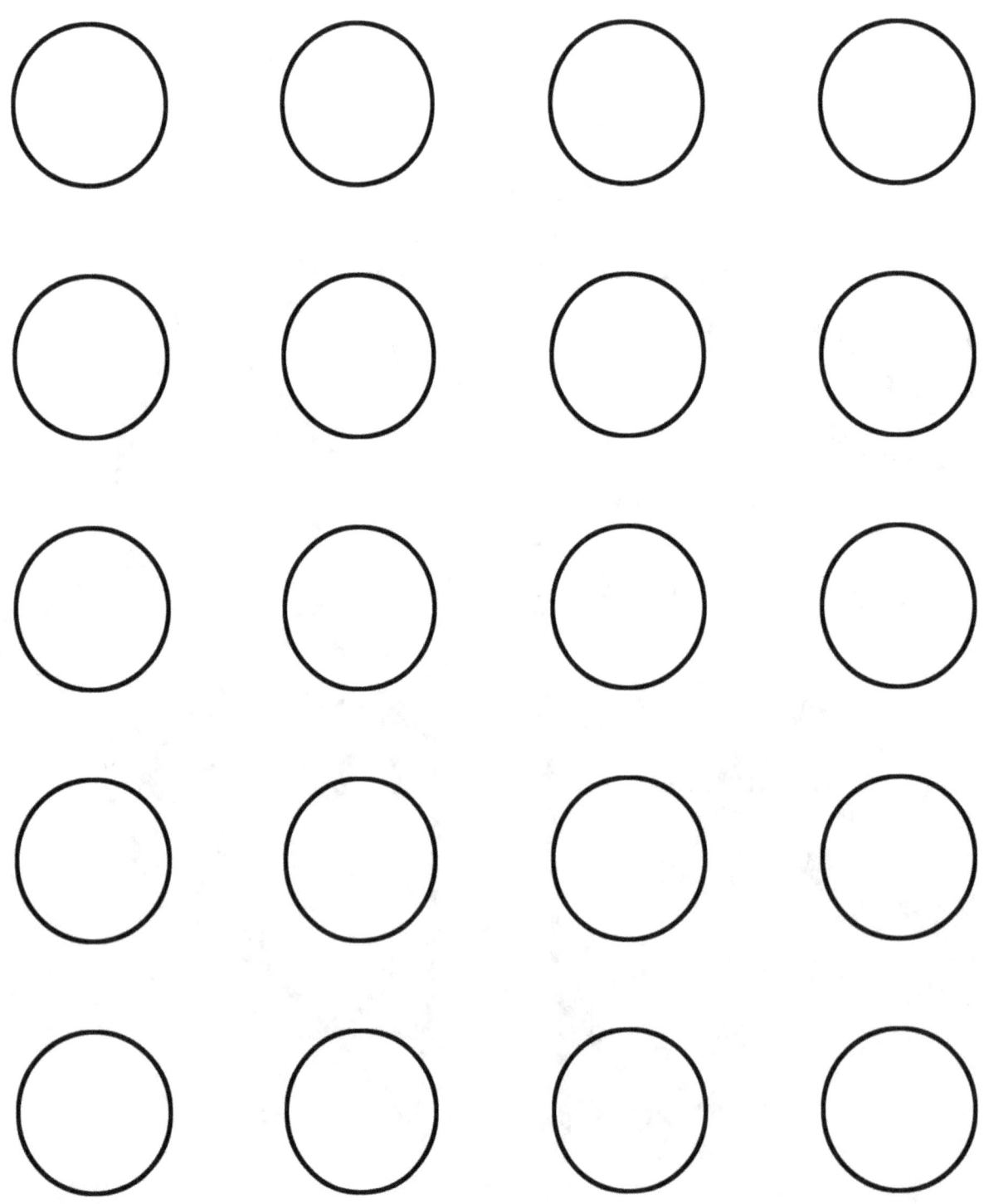

Continue to join the dots you have connected all the numbered dots. Then, color the picture!

Test Your Color

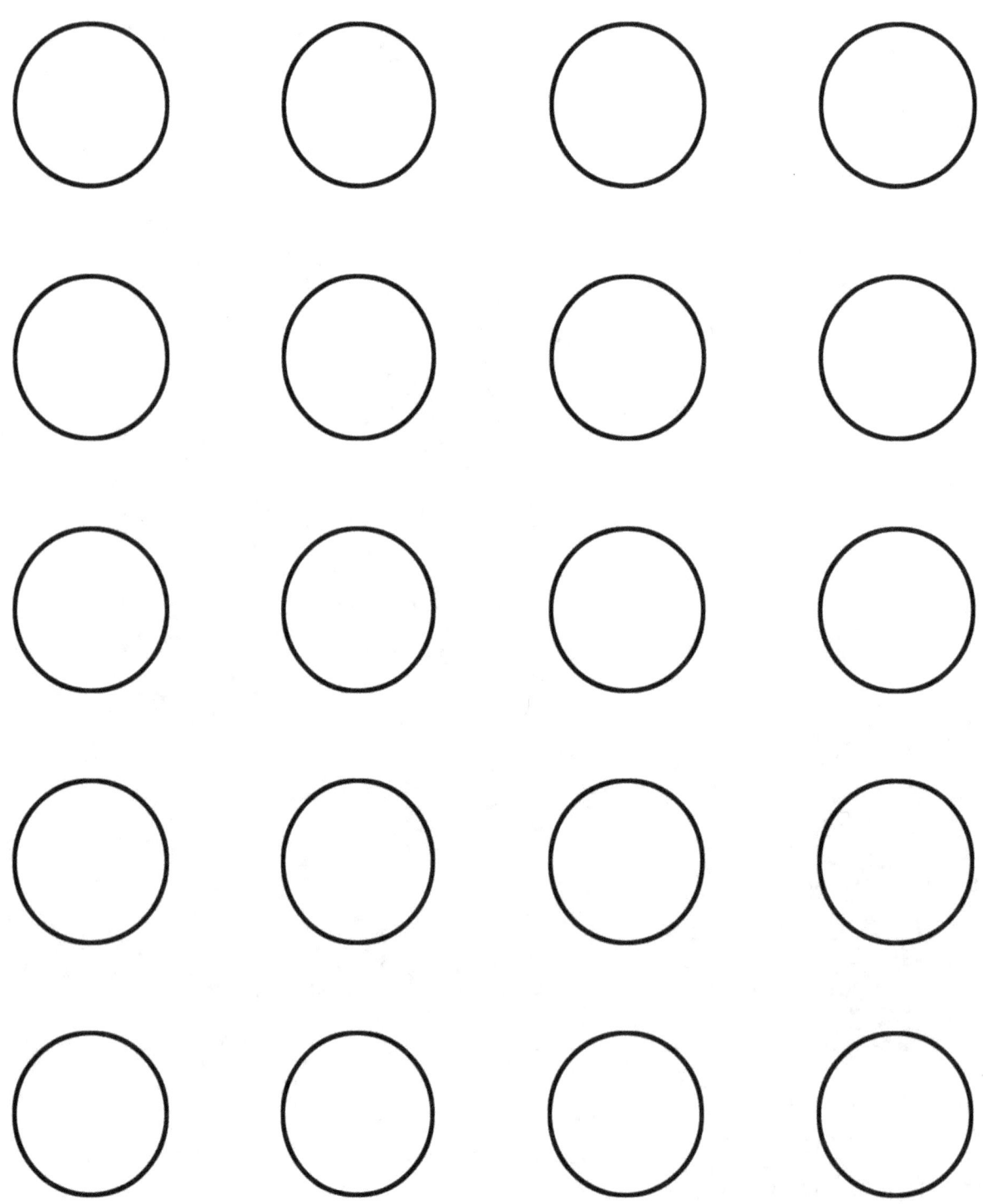

Continue to join the dots you have connected all the numbered dots. Then, color the picture!

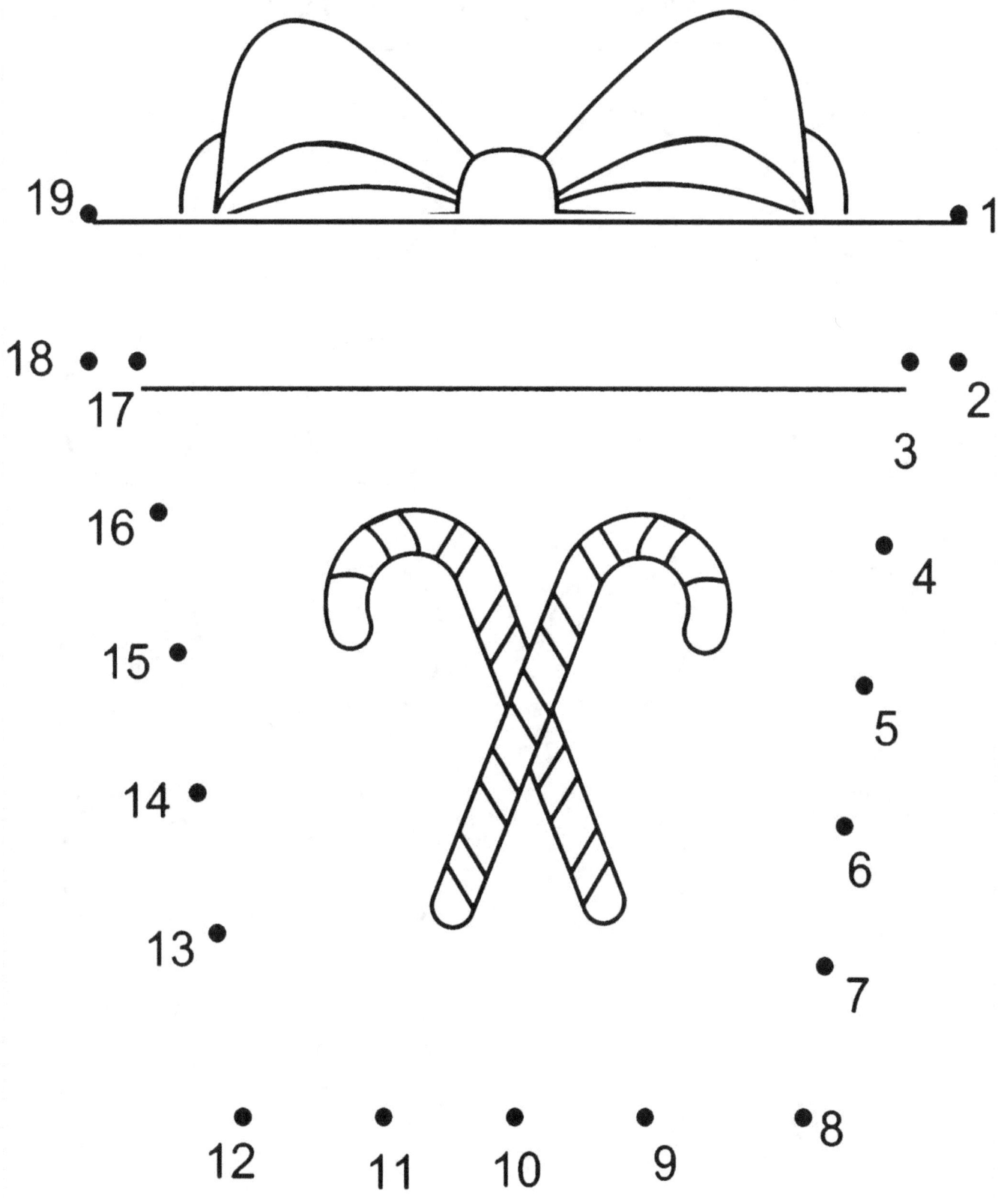

Test Your Color

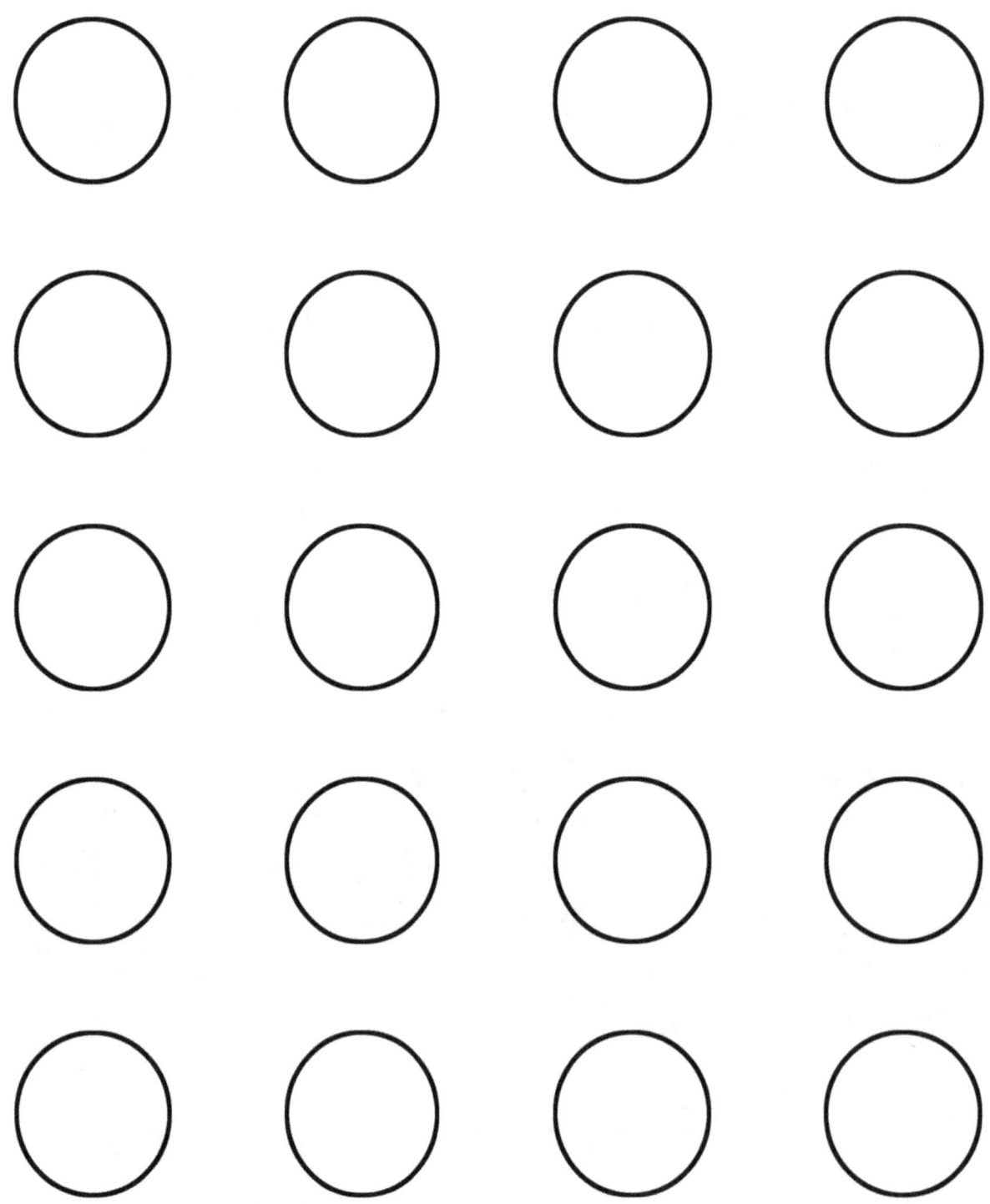

Continue to join the dots you have connected all the numbered dots. Then, color the picture!

Test Your Color

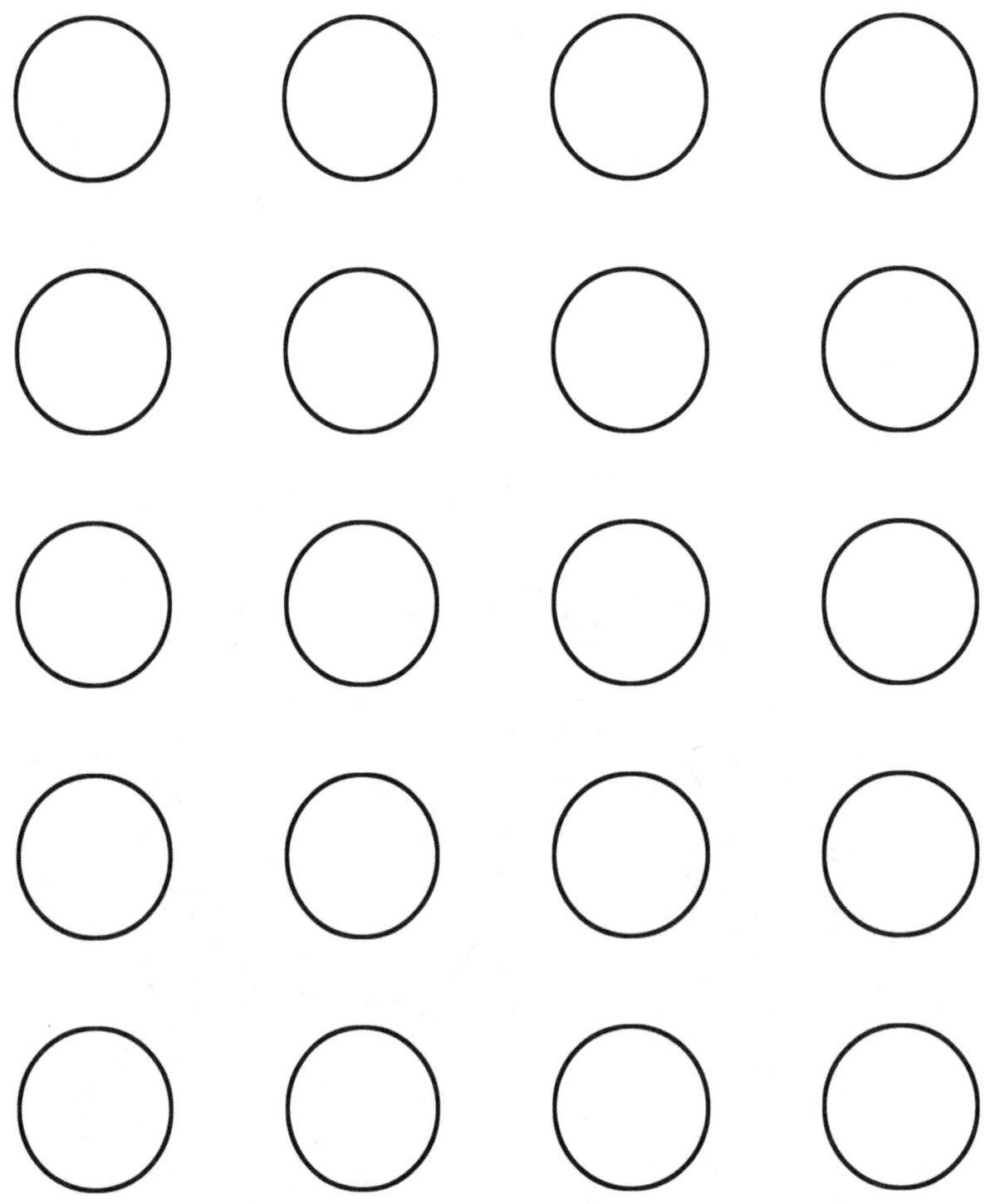

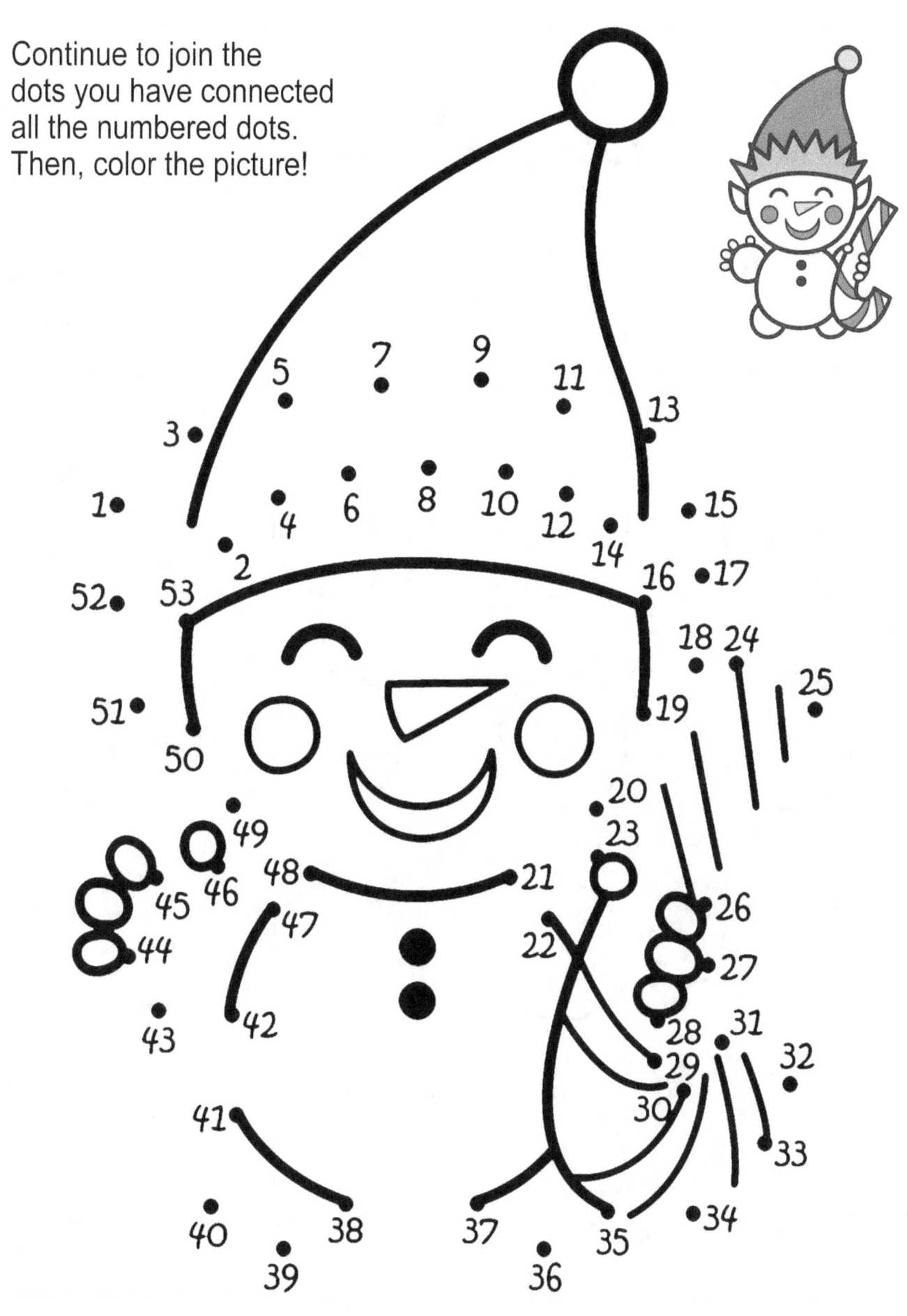

Test Your Color

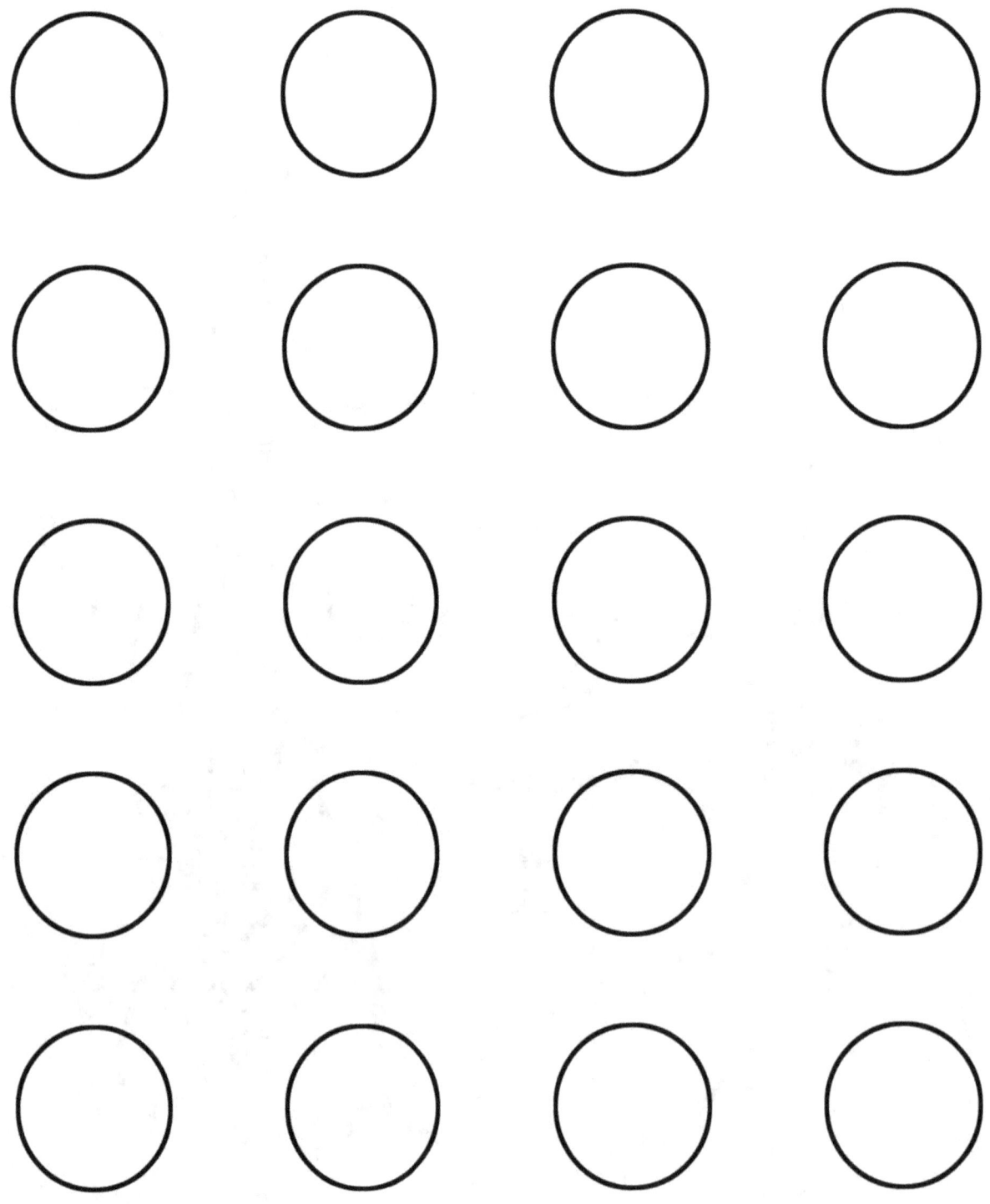

Continue to join the dots you have connected all the numbered dots. Then, color the picture!

Test Your Color

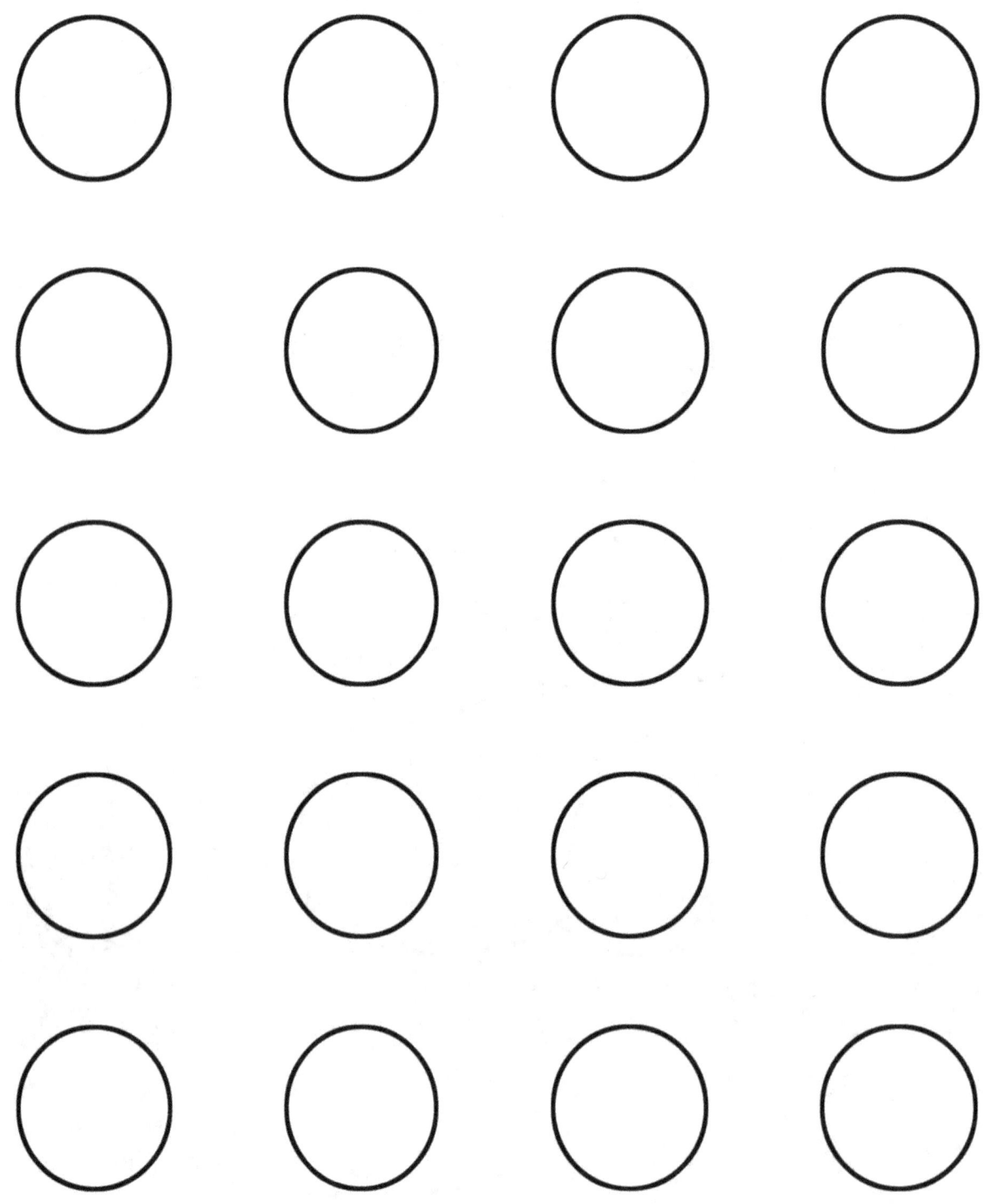

Continue to join the dots you have connected all the numbered dots. Then, color the picture!

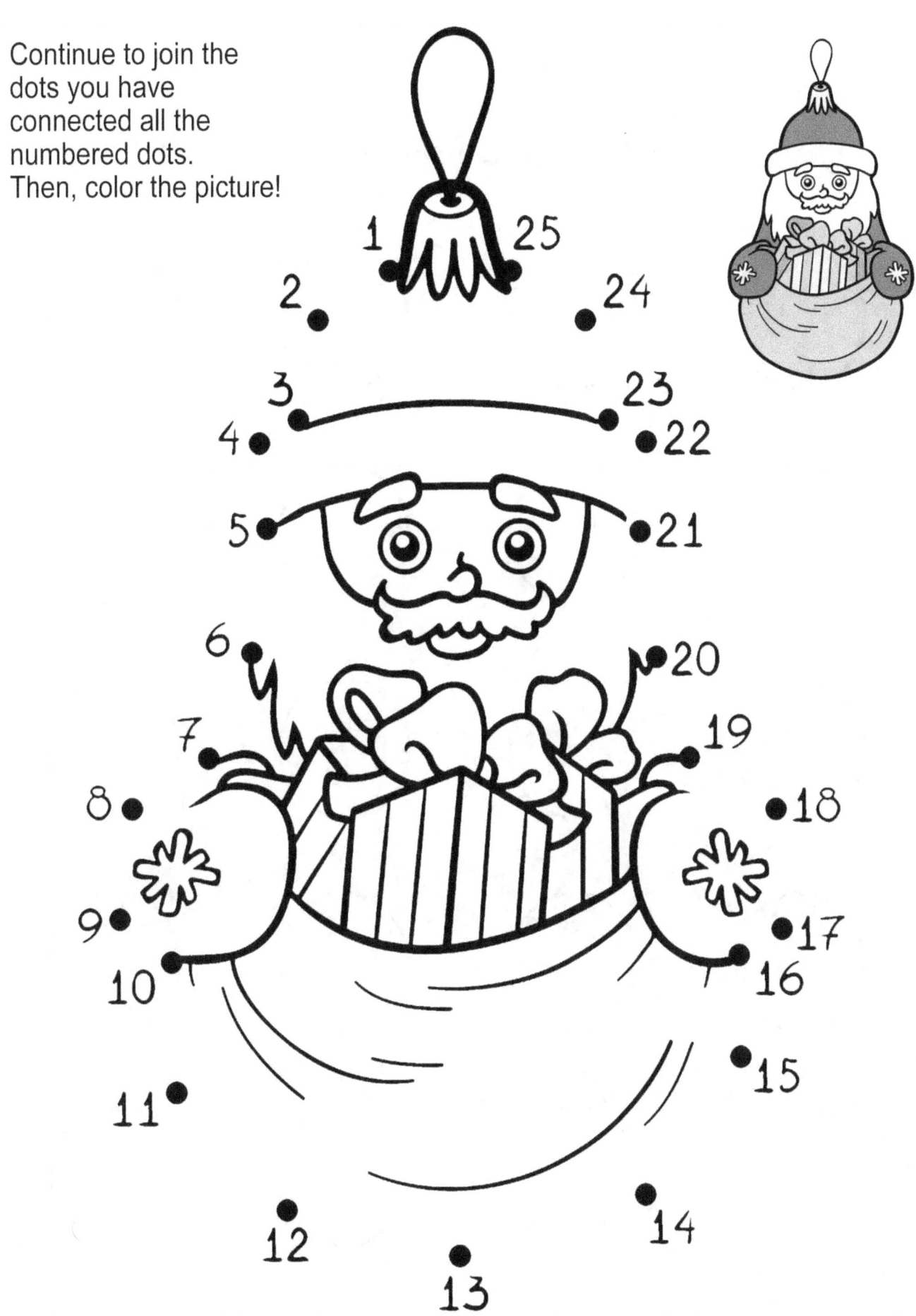

Test Your Color

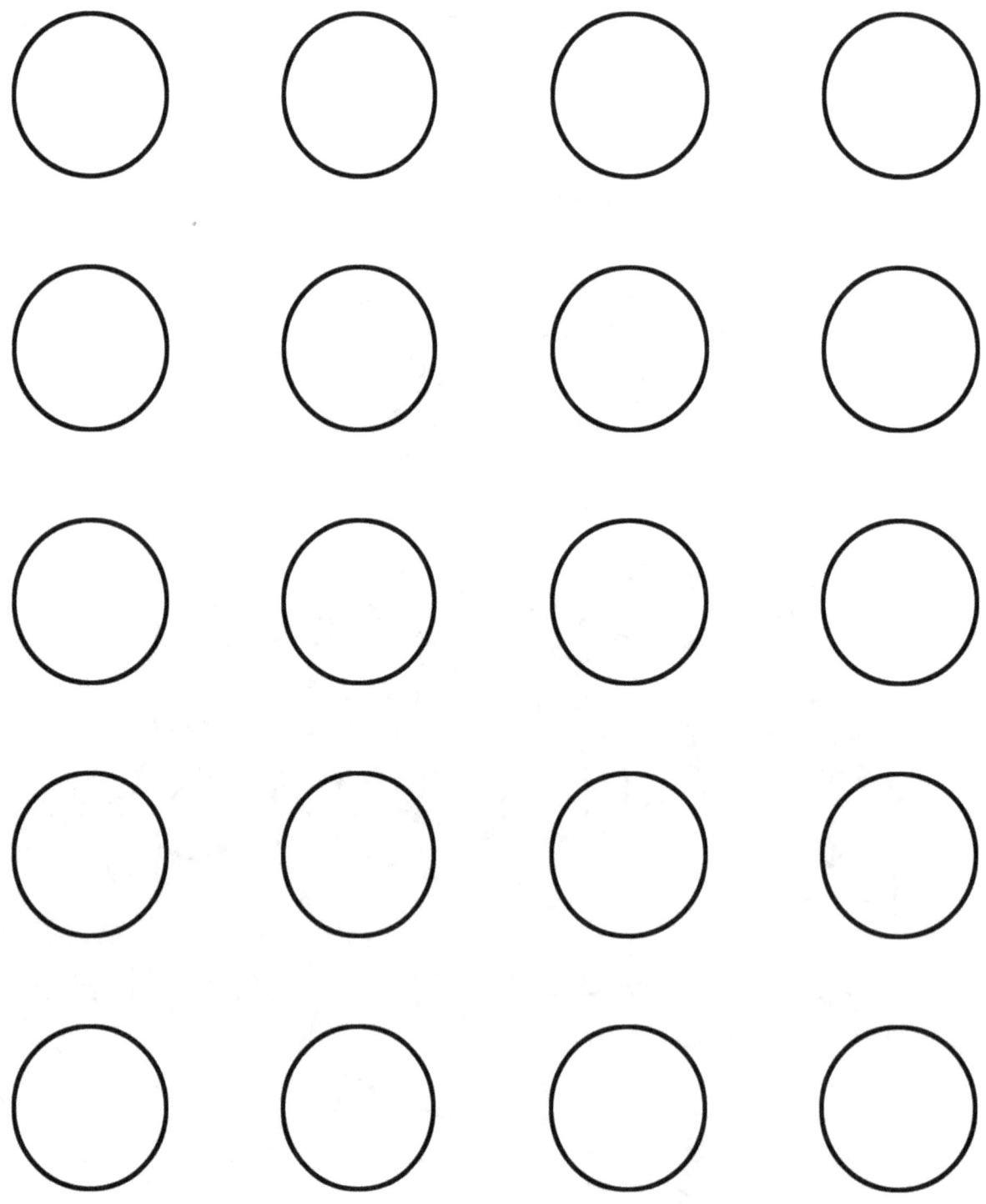

Test Your Color

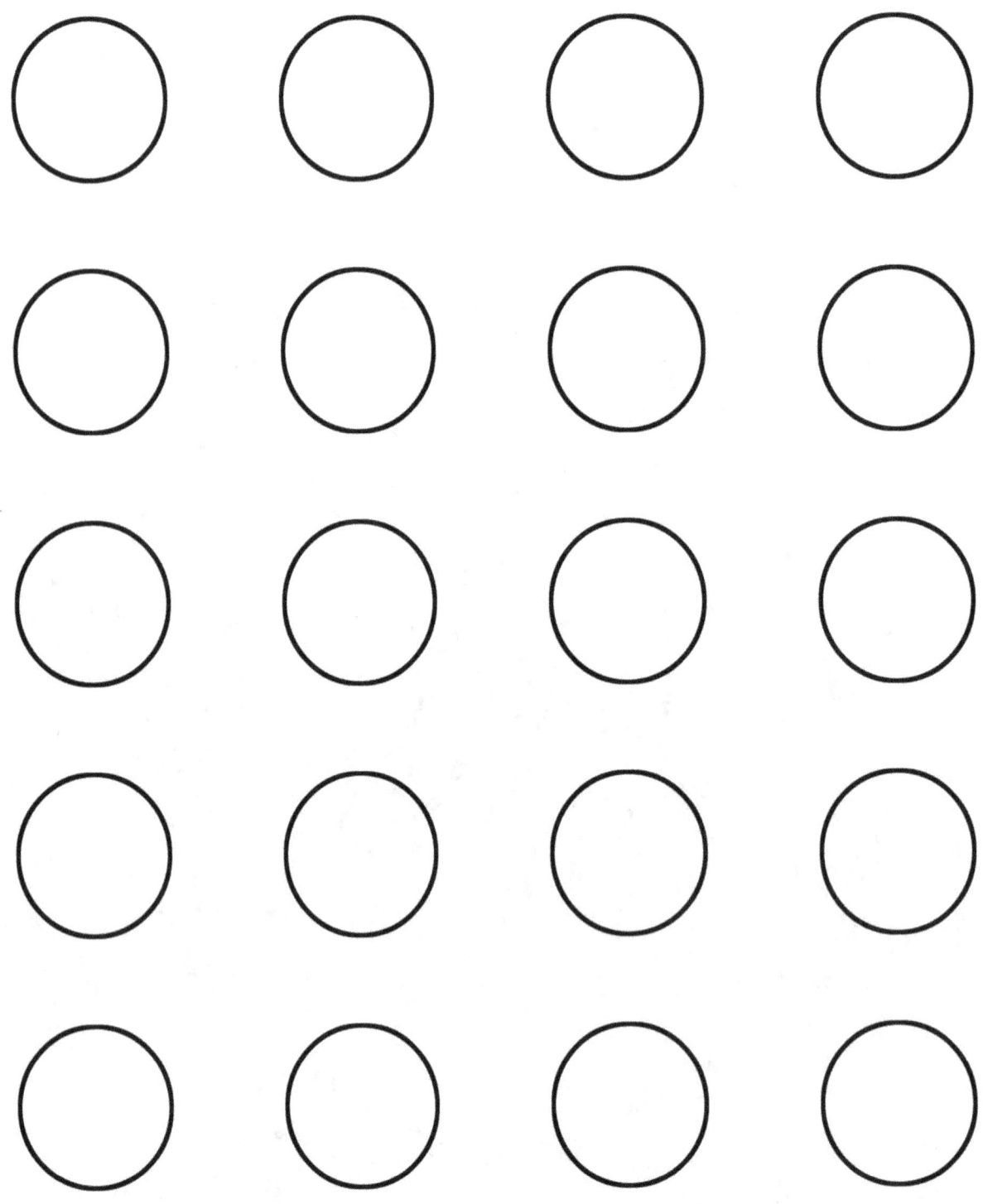

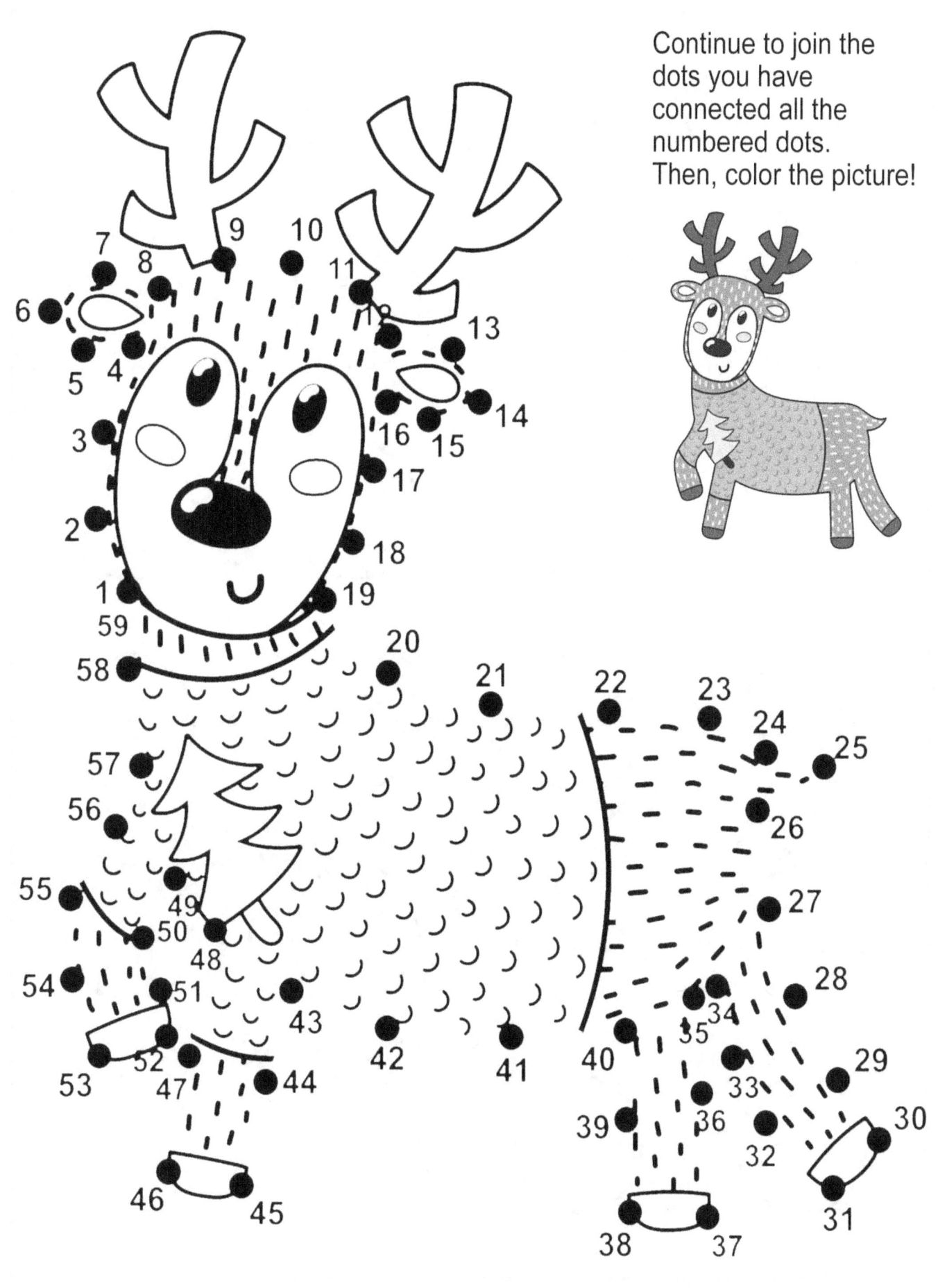

Test Your Color

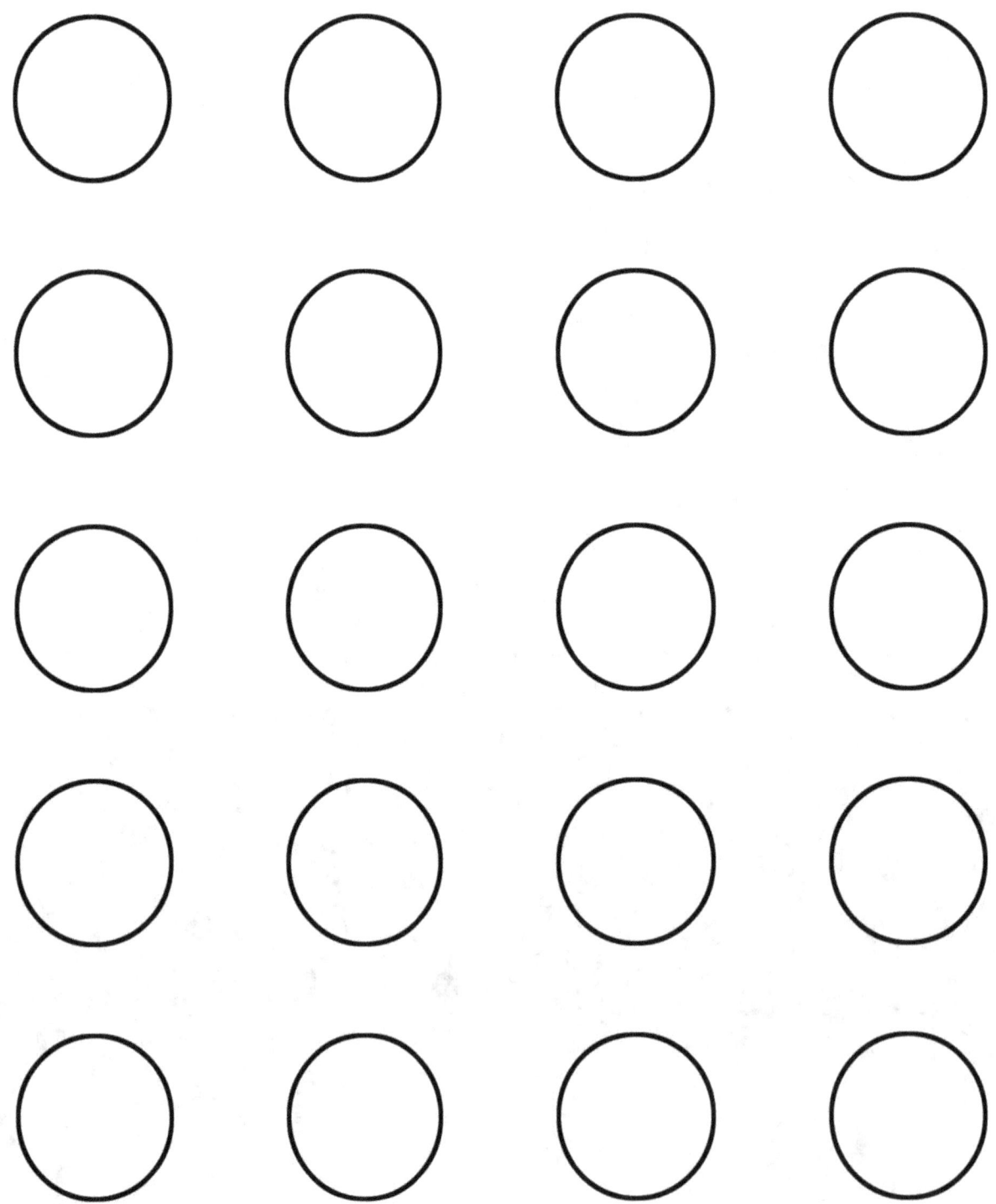

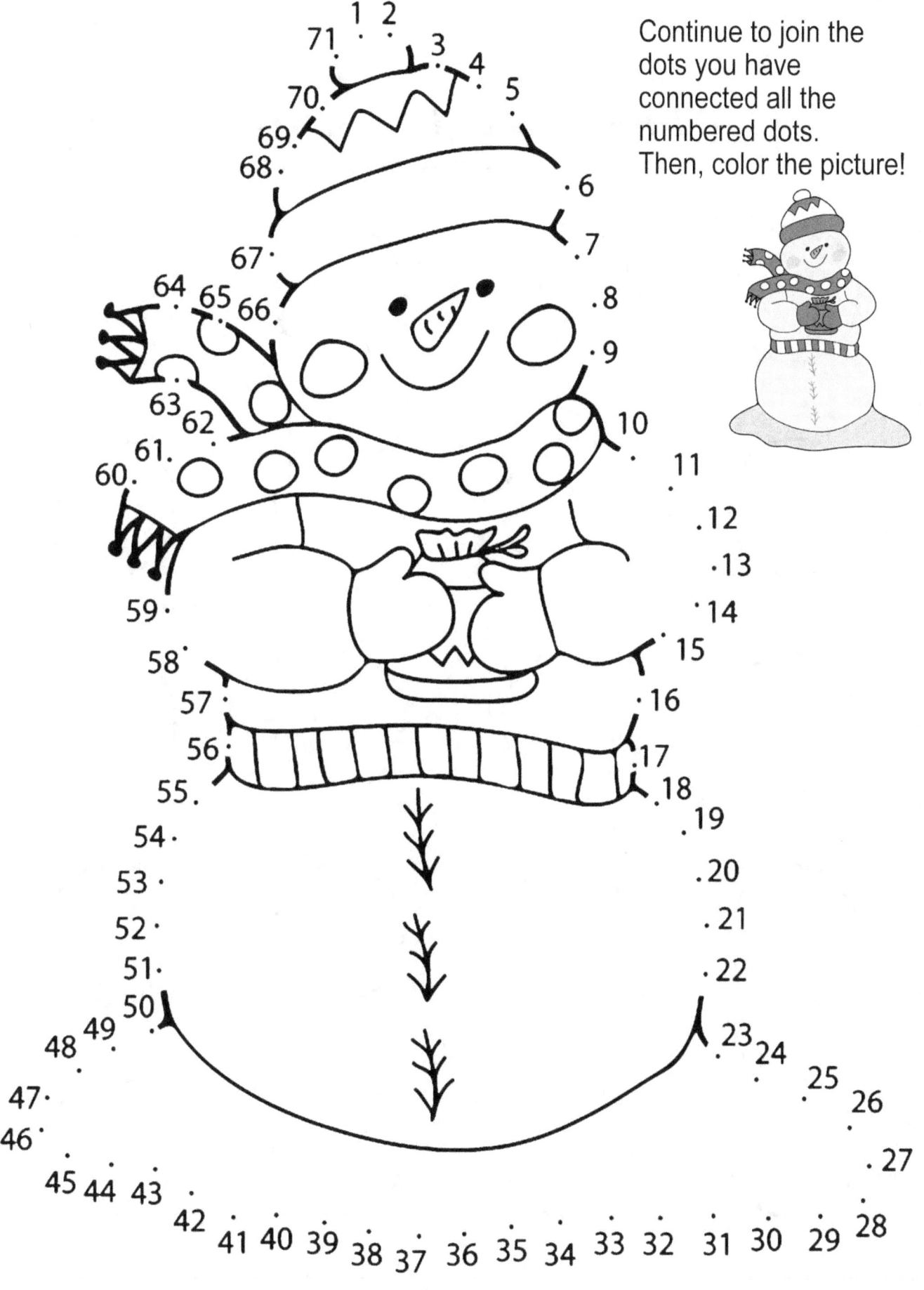

Continue to join the dots you have connected all the numbered dots. Then, color the picture!

Test Your Color

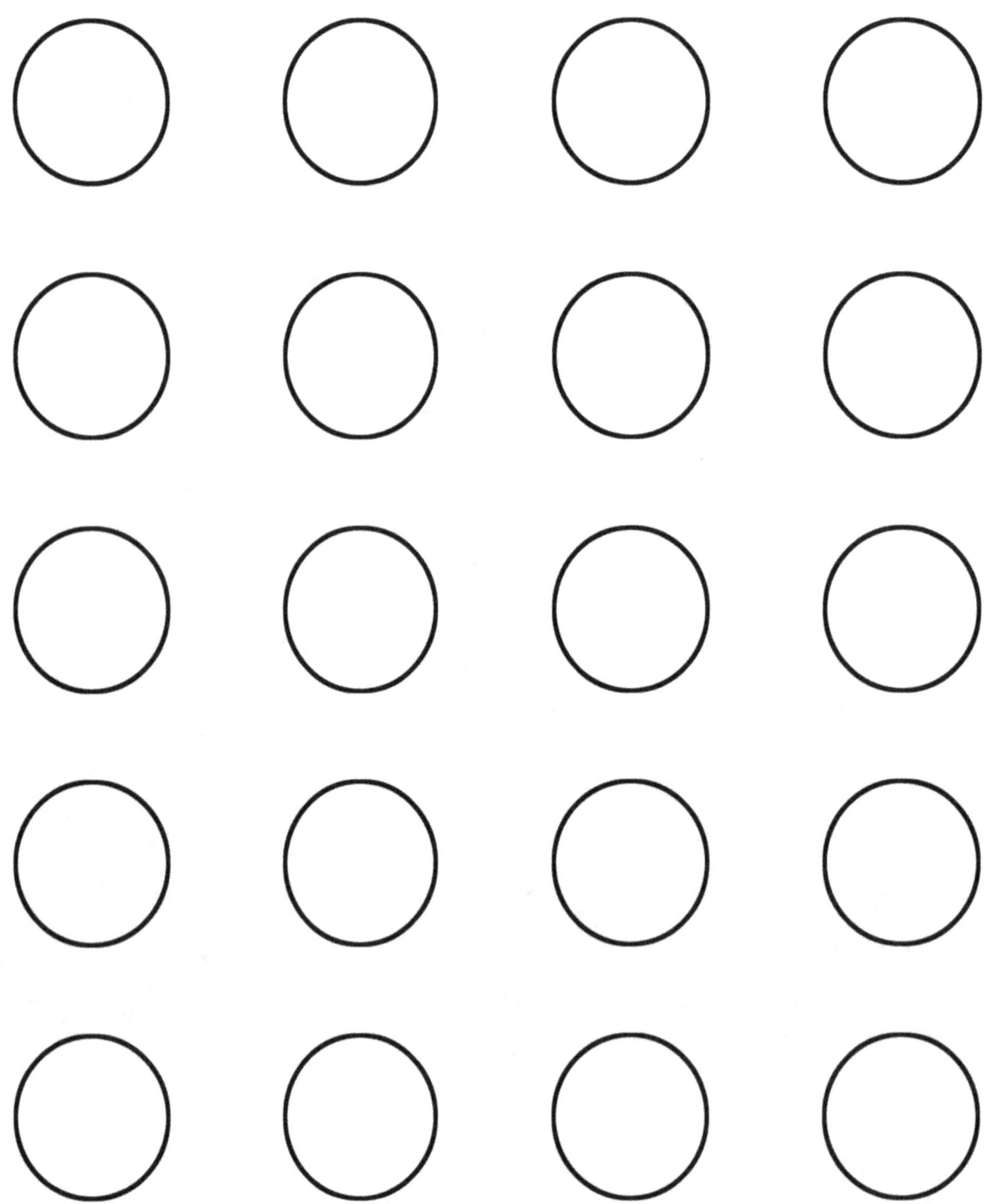

Continue to join the dots you have connected all the numbered dots. Then, color the picture!

Test Your Color

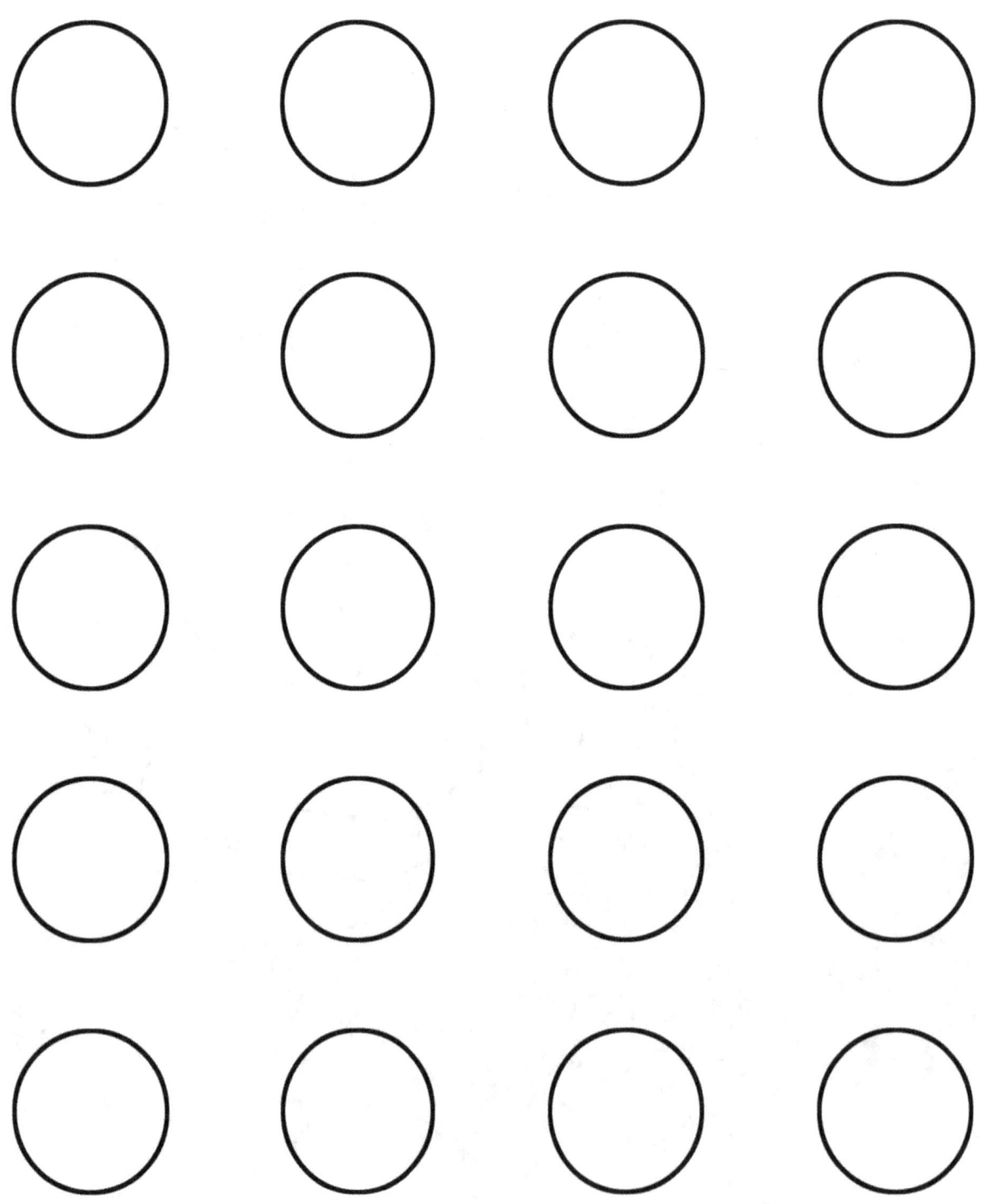

Continue to join the dots you have connected all the numbered dots. Then, color the picture!

Test Your Color

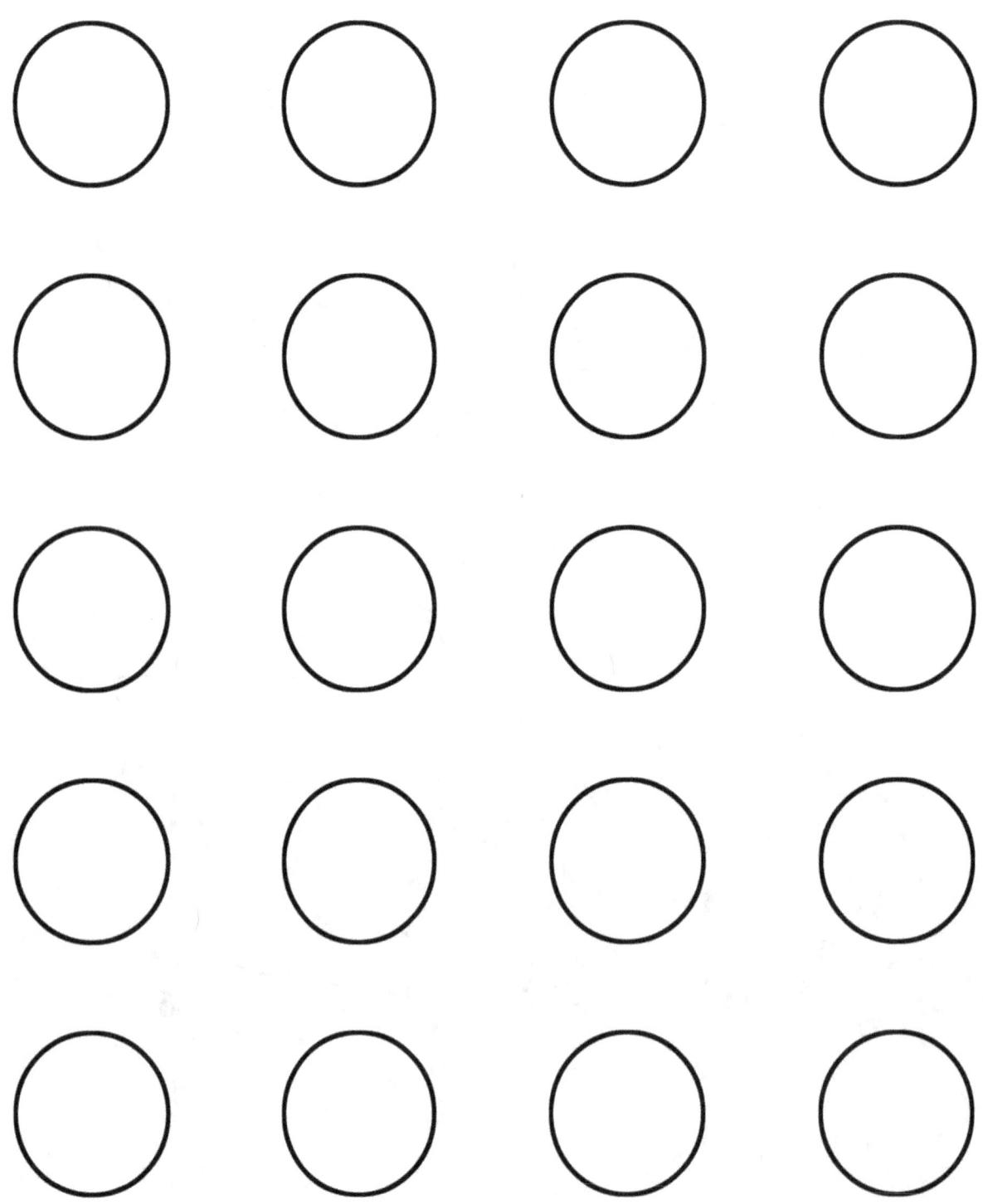

Continue to join the dots you have connected all the numbered dots. Then, color the picture!

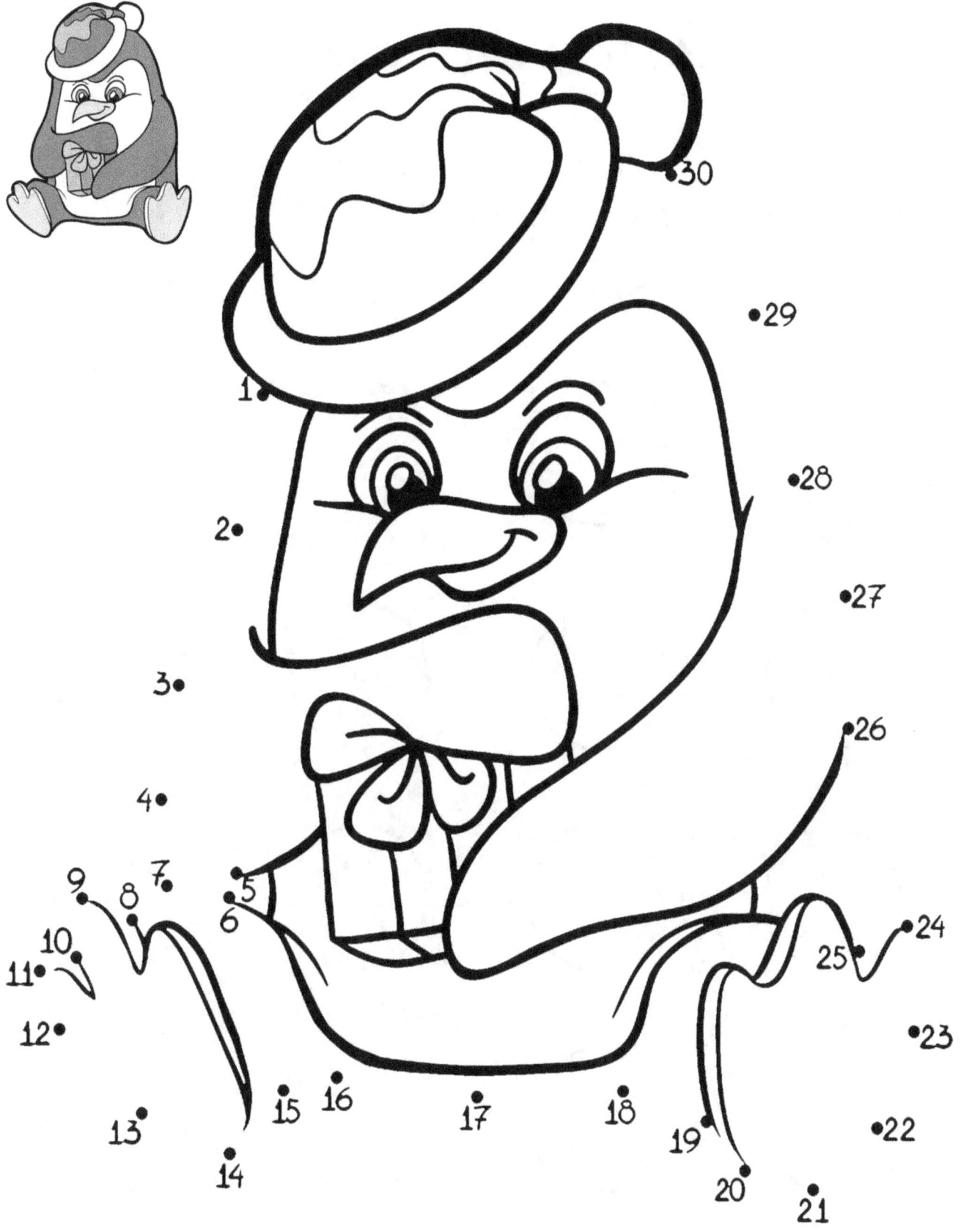

Test Your Color

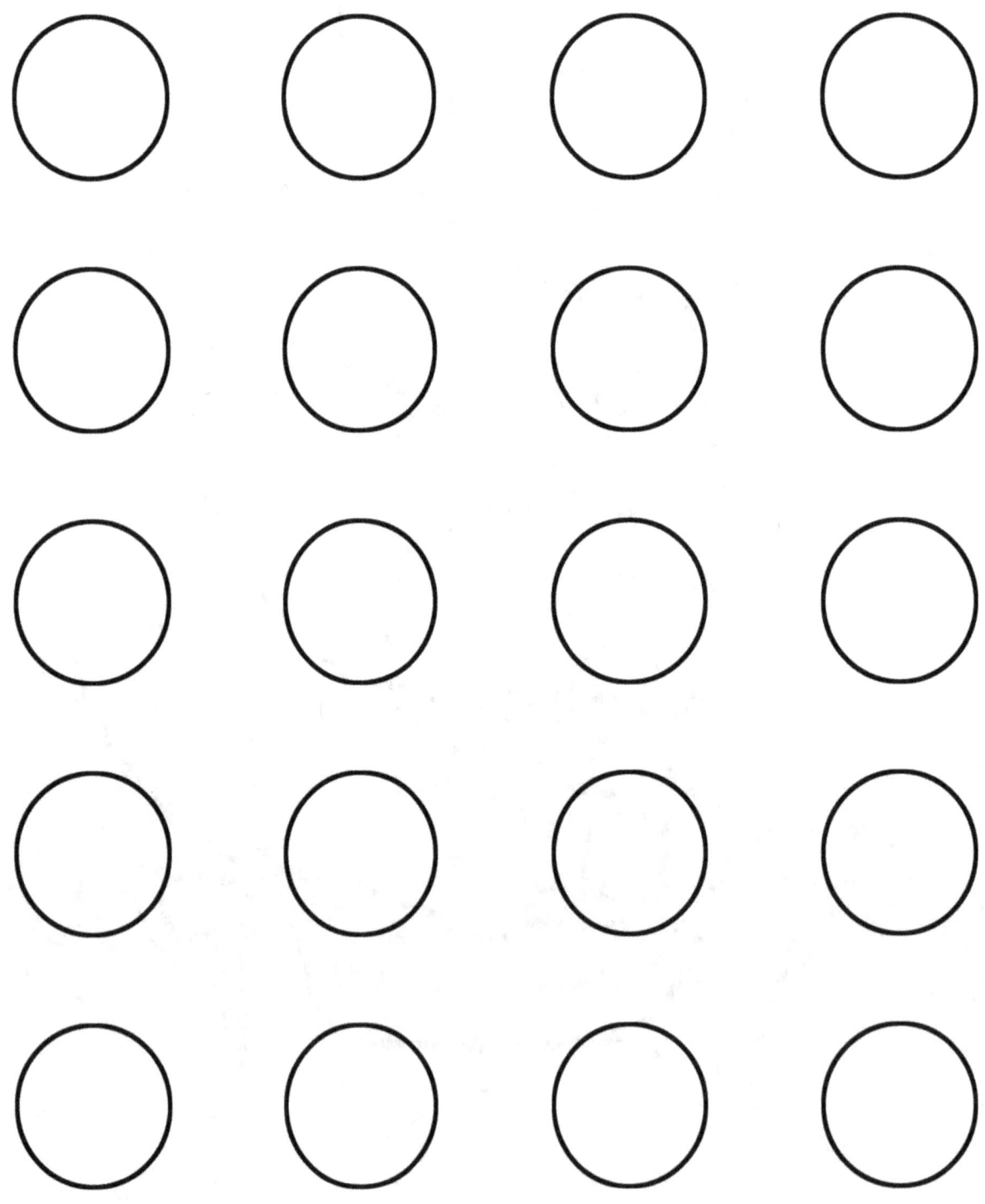

Continue to join the dots you have connected all the numbered dots. Then, color the picture!

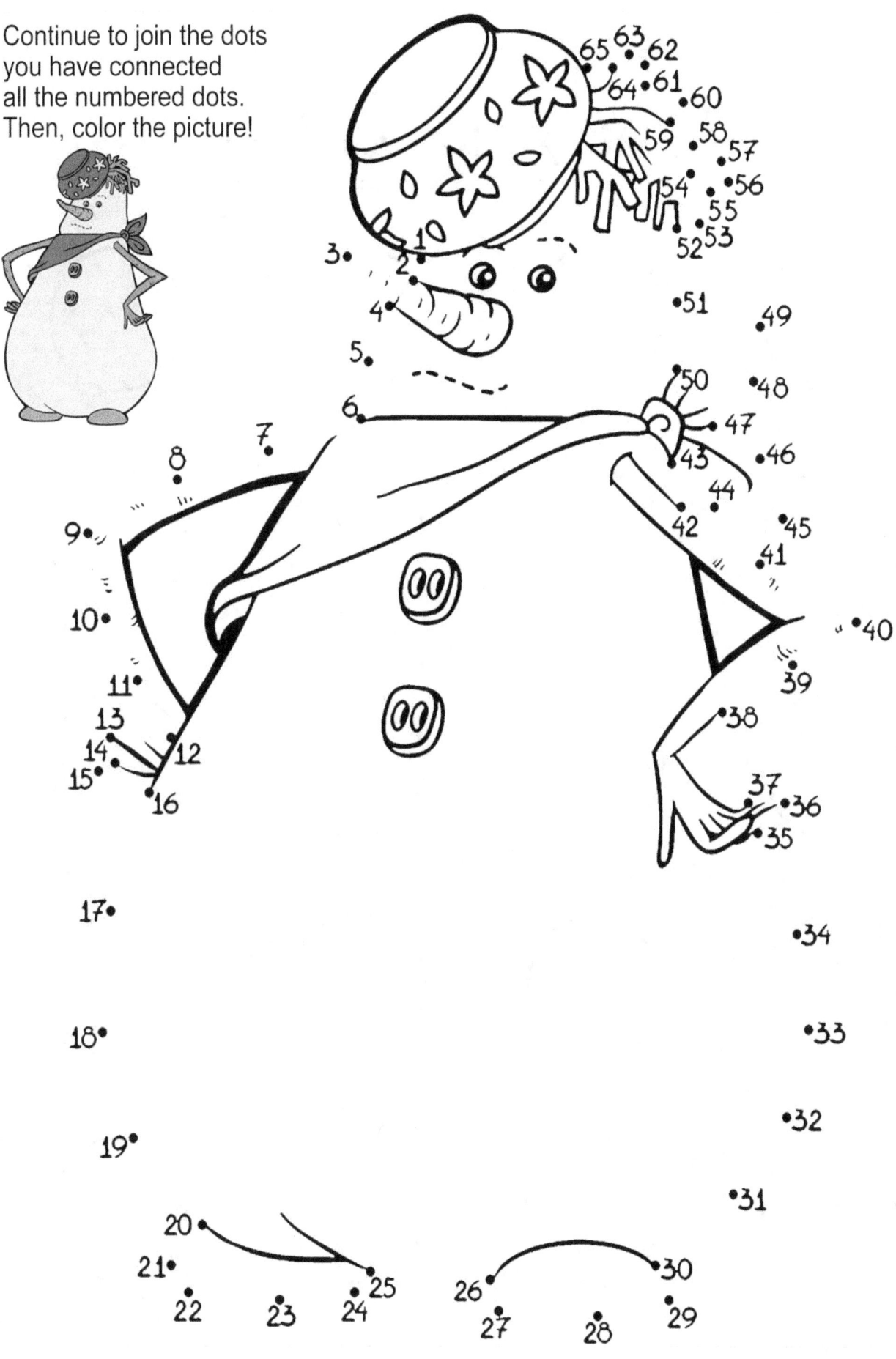

Test Your Color

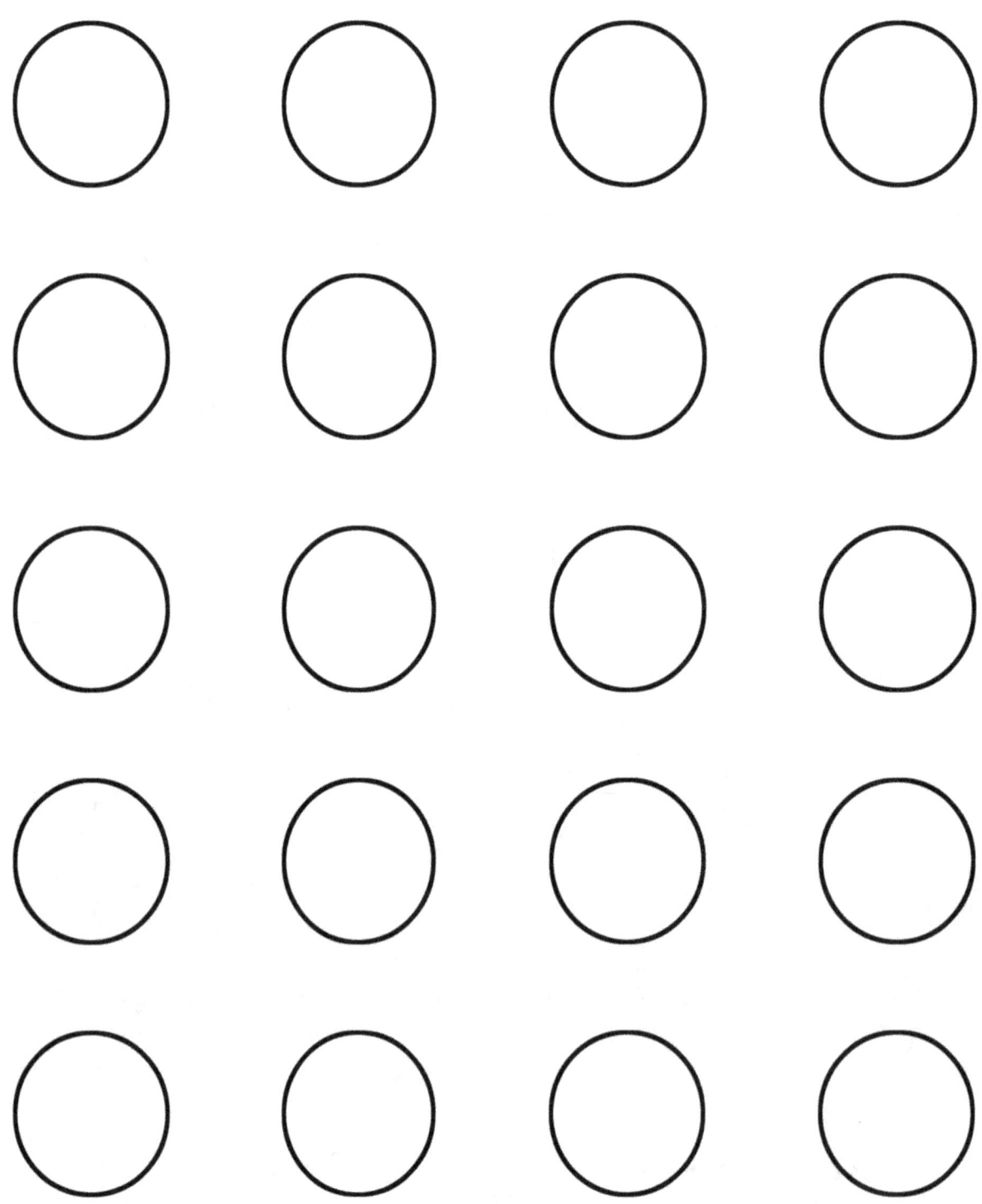

Test Your Color

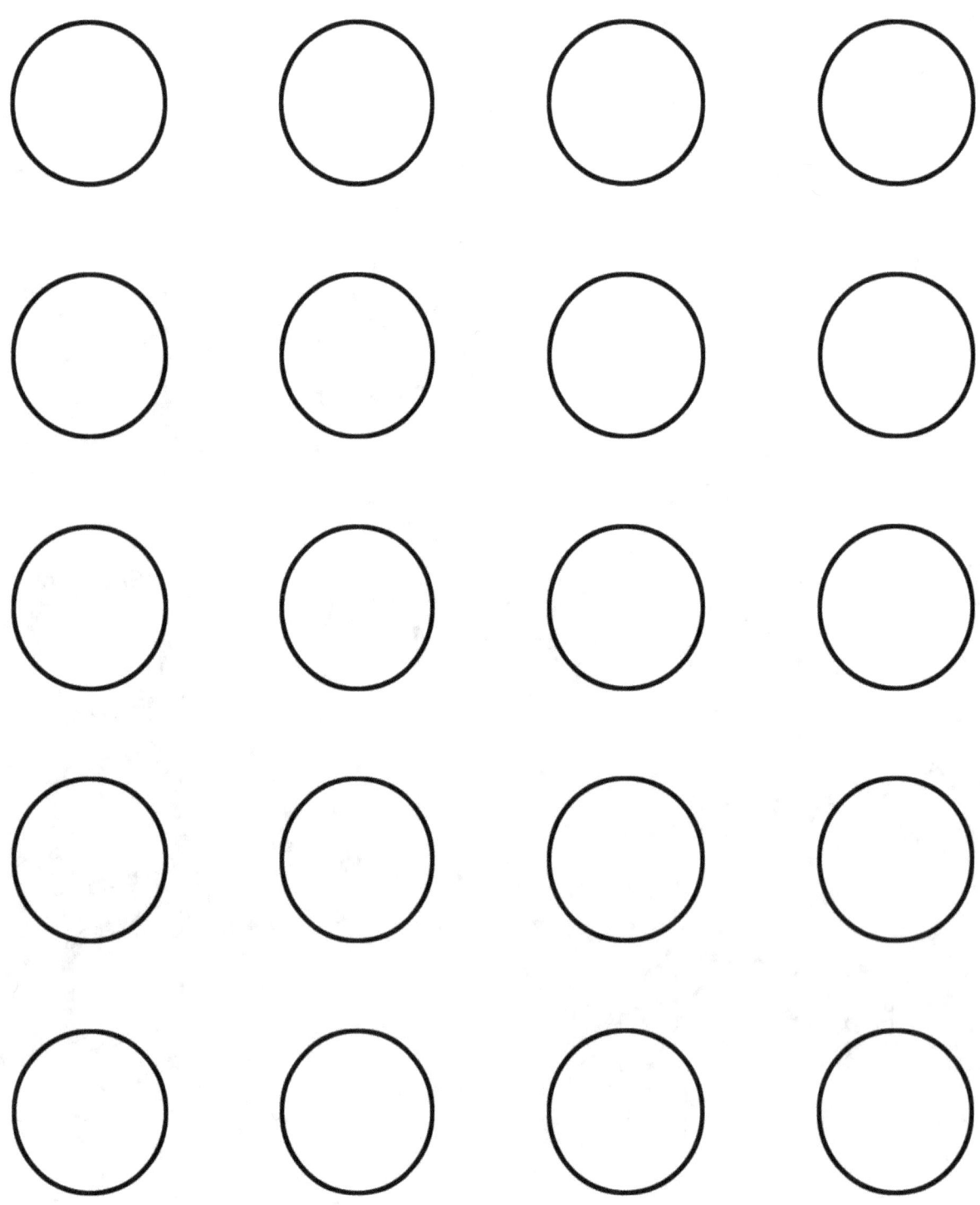

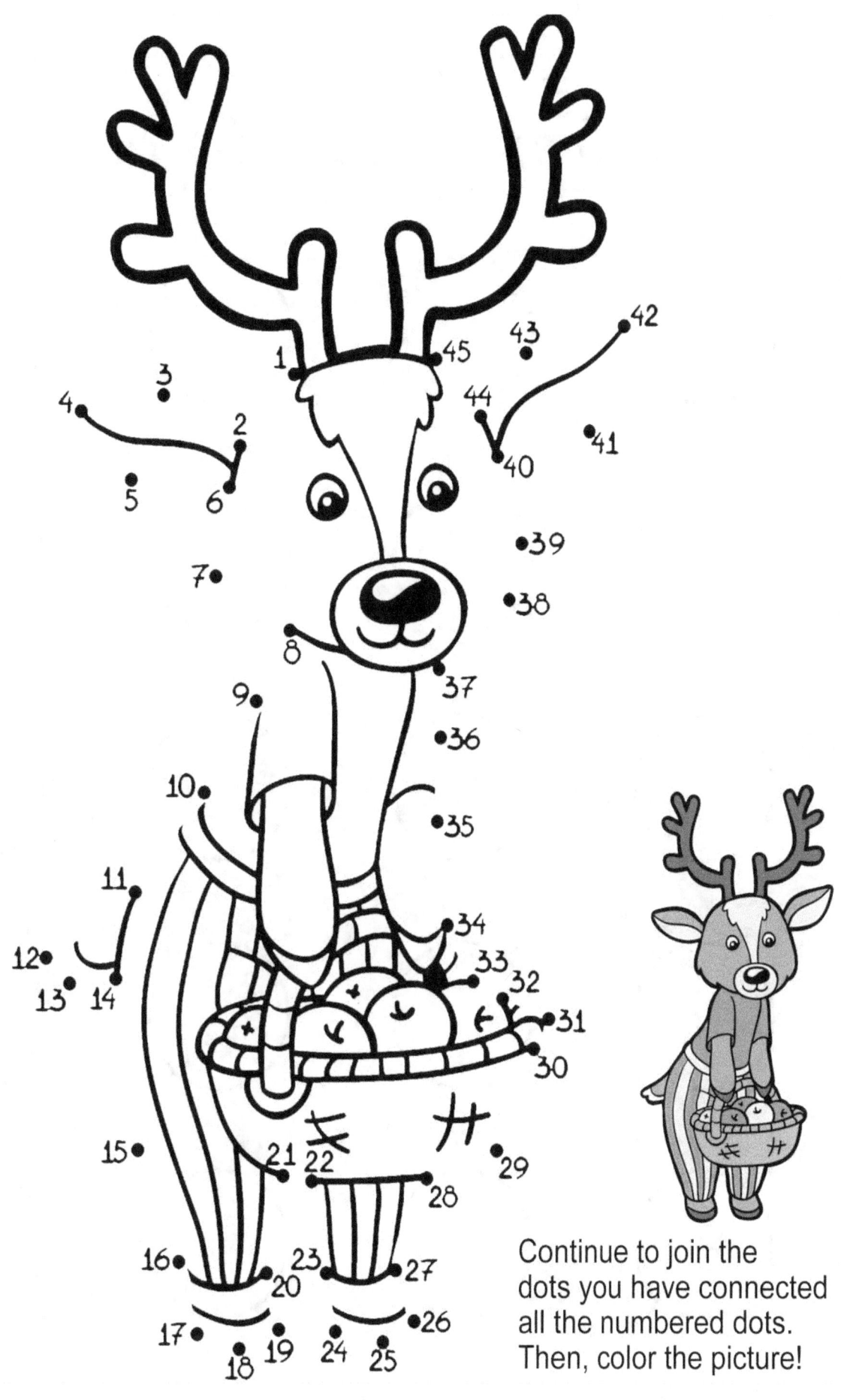

Continue to join the dots you have connected all the numbered dots. Then, color the picture!

Test Your Color

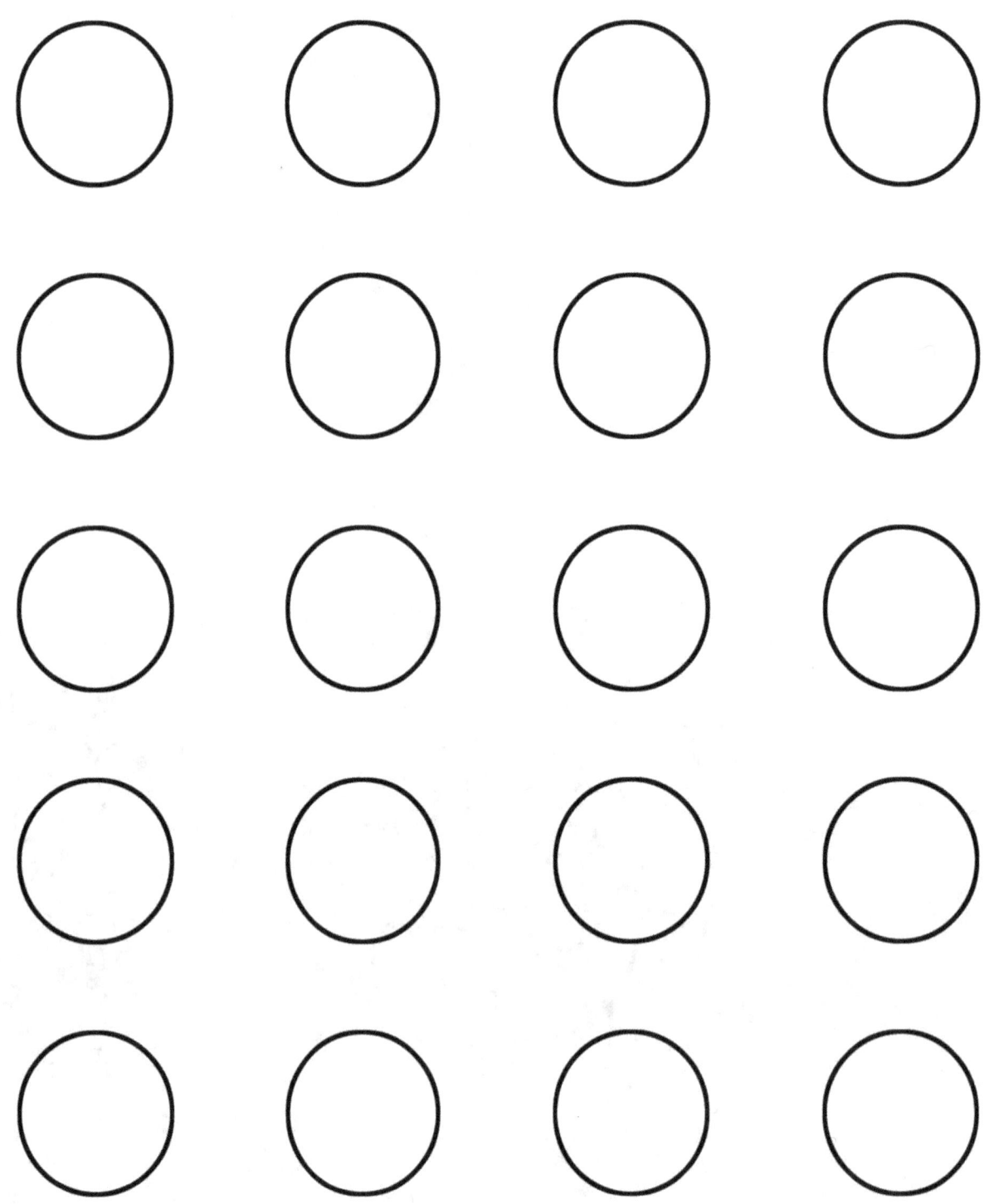

Continue to join the dots you have connected all the numbered dots. Then, color the picture!

Test Your Color

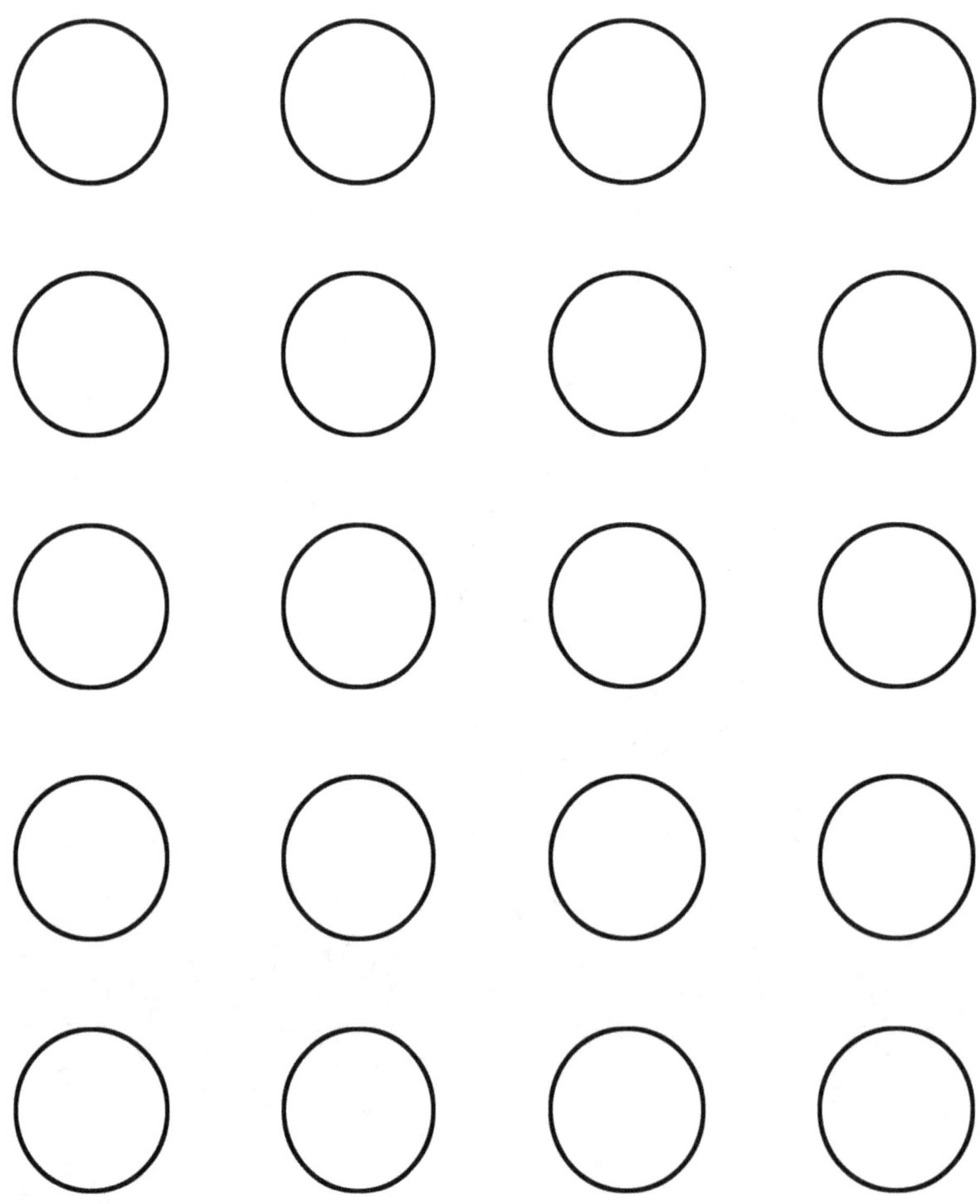

Continue to join the dots you have connected all the numbered dots. Then, color the picture!

Test Your Color

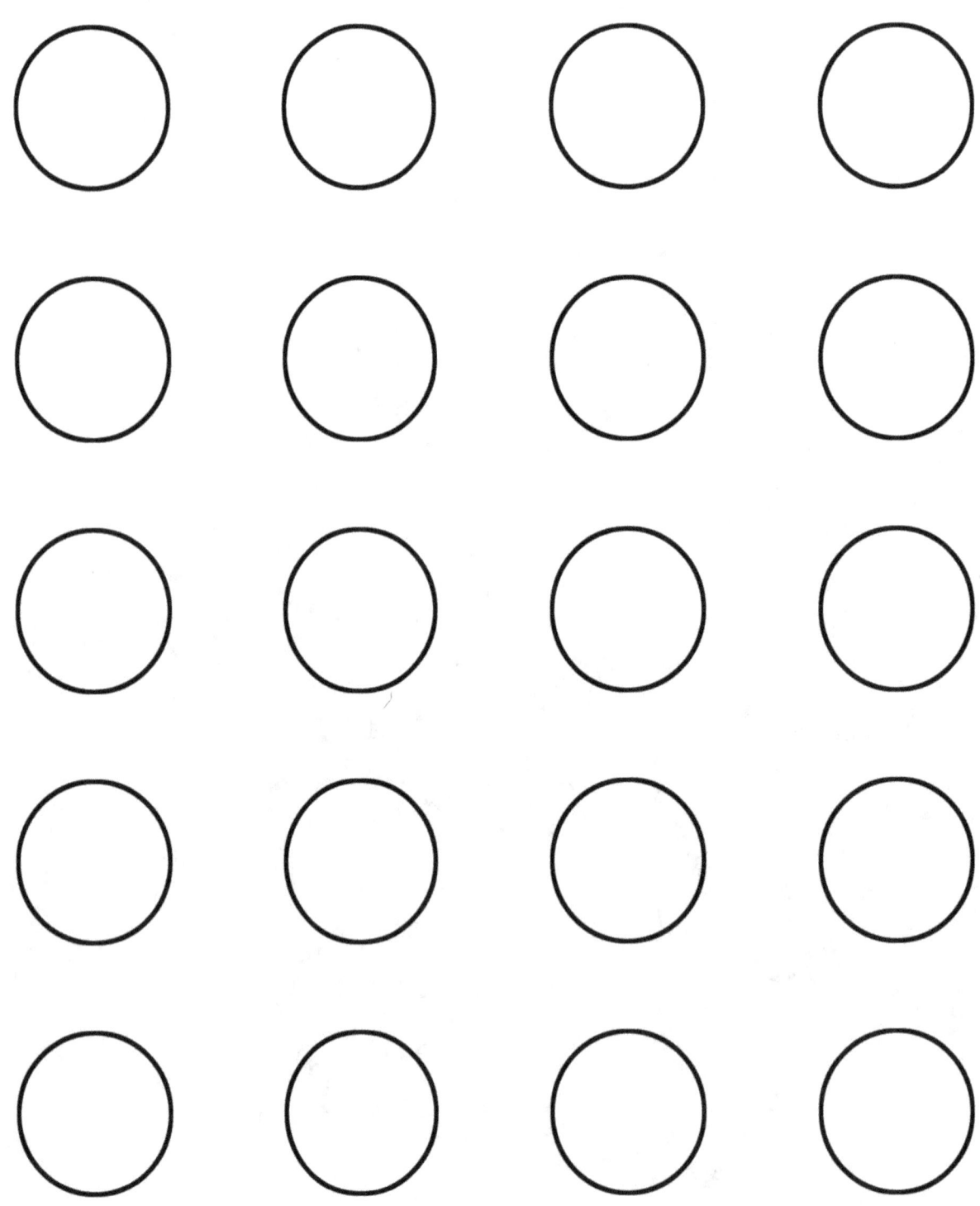

Continue to join the dots you have connected all the numbered dots. Then, color the picture!

Test Your Color

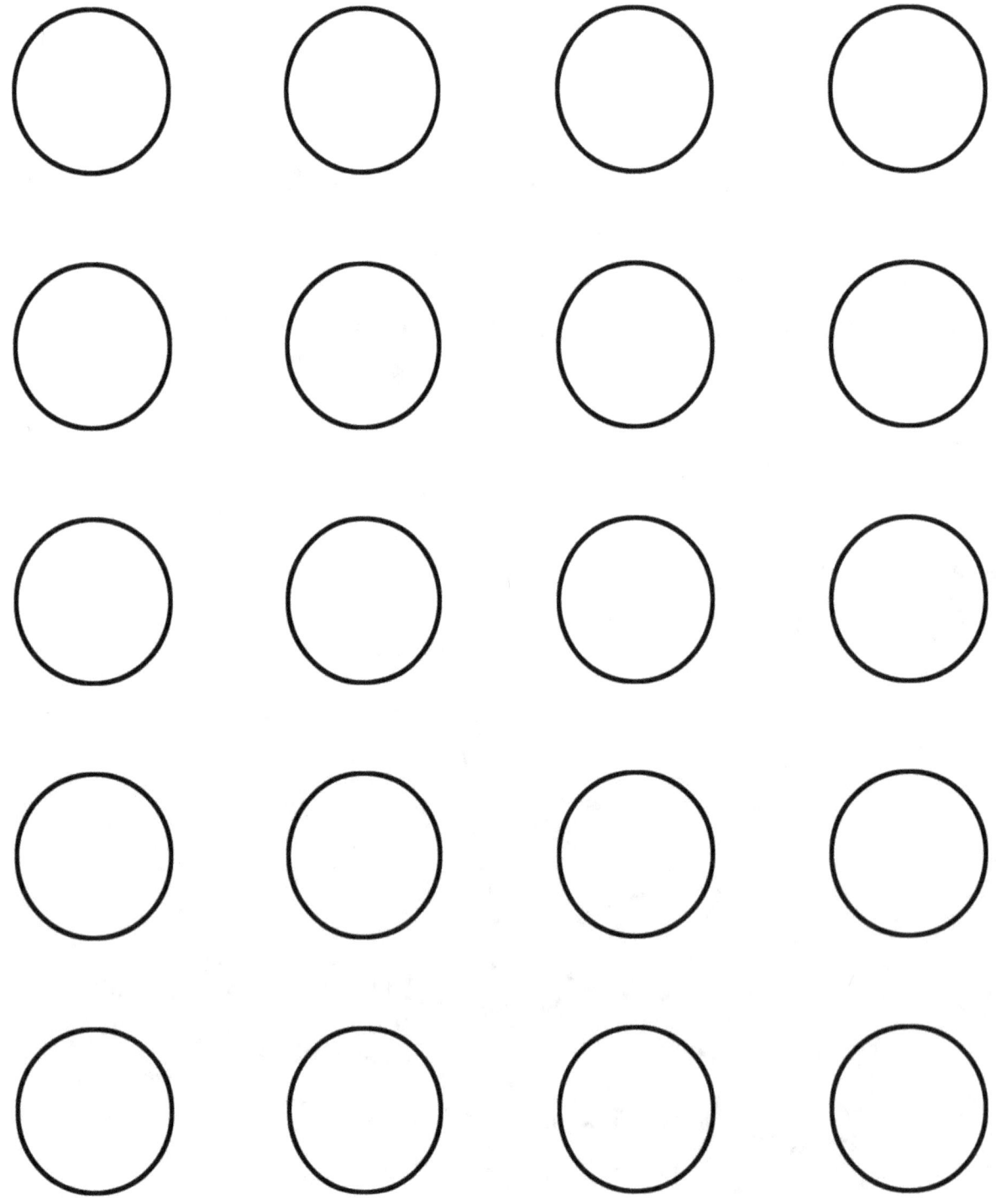

Continue to join the dots you have connected all the numbered dots. Then, color the picture!

Test Your Color

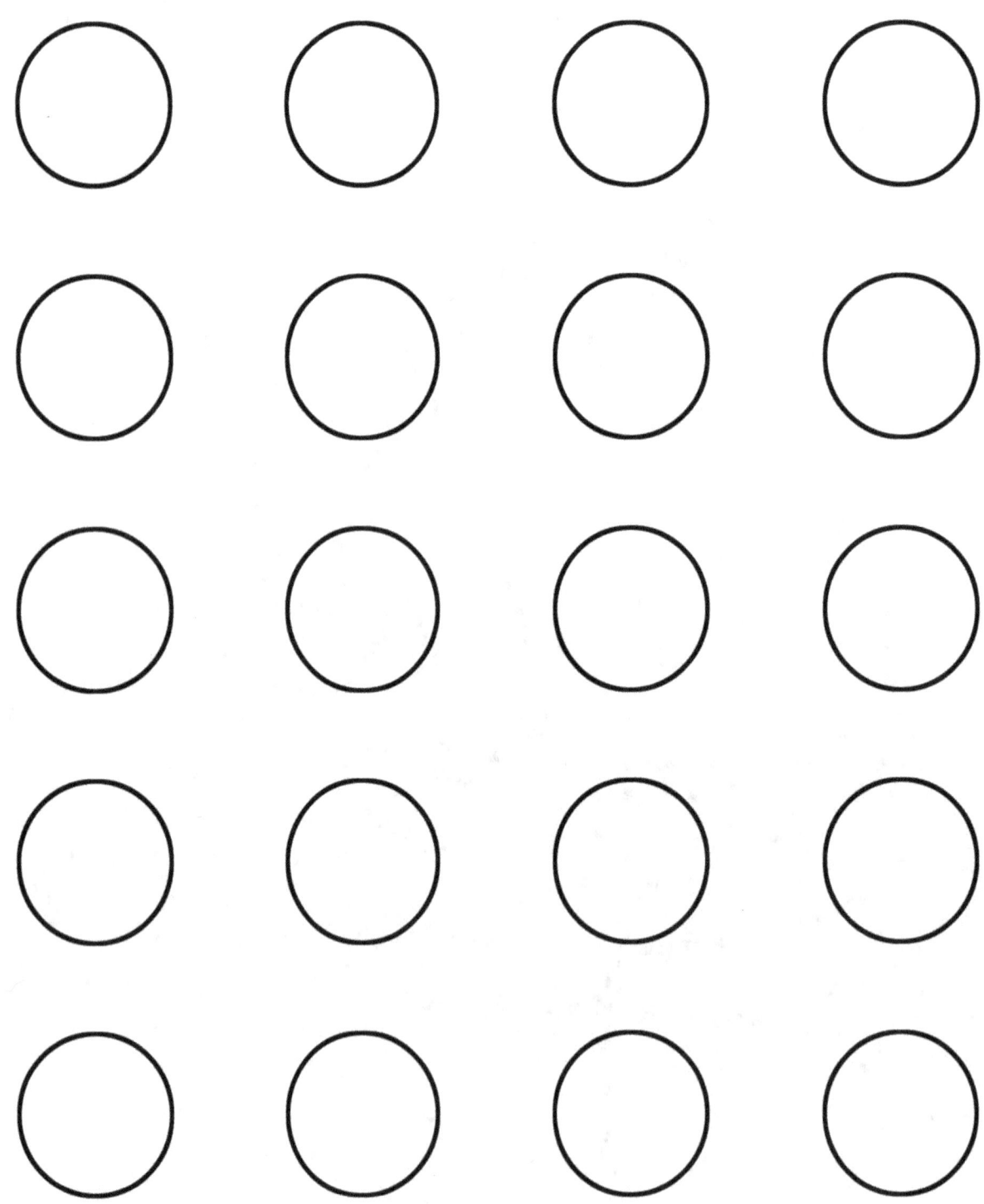

Test Your Color

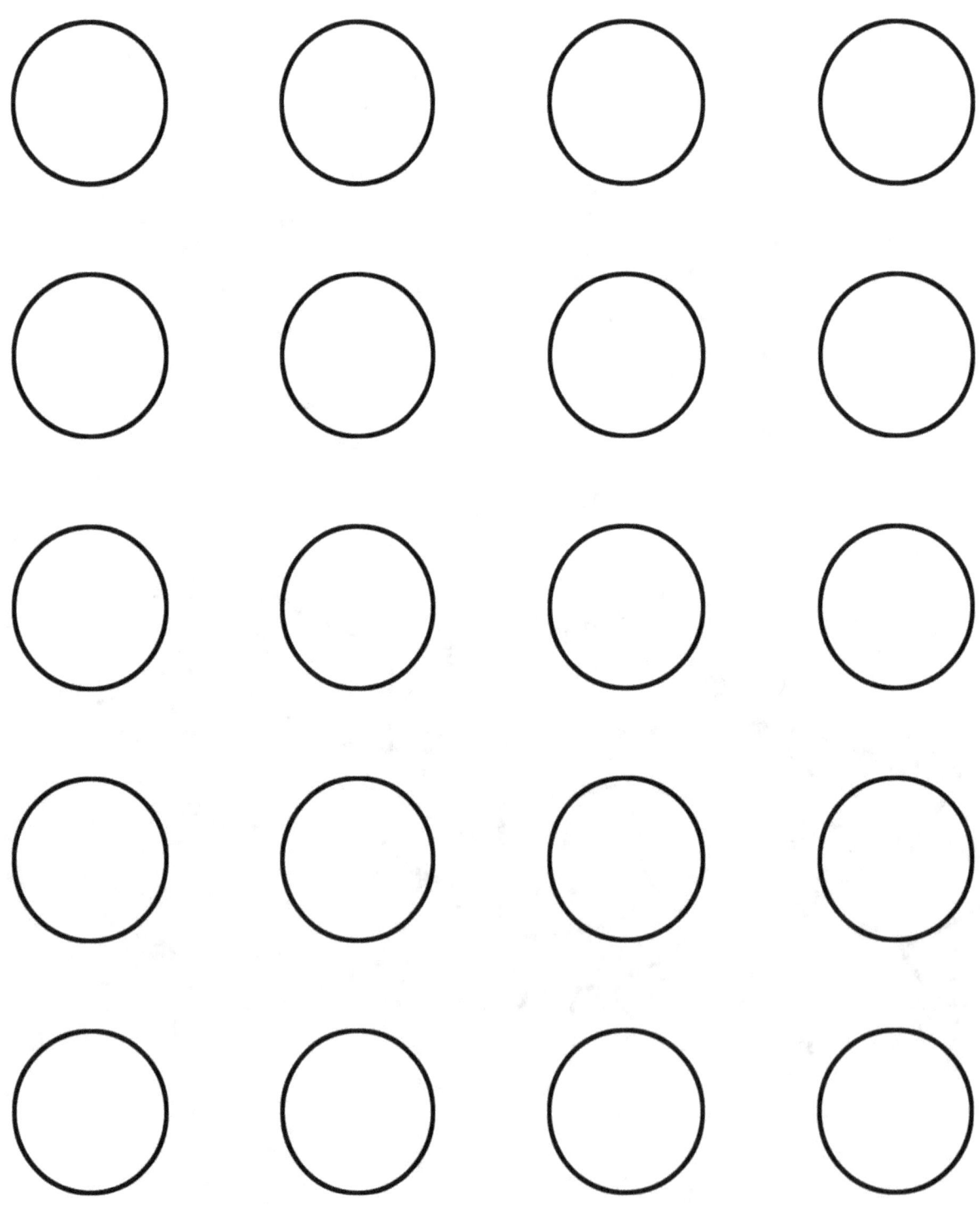

Continue to join the dots you have connected all the numbered dots.
Then, color the picture!

Test Your Color

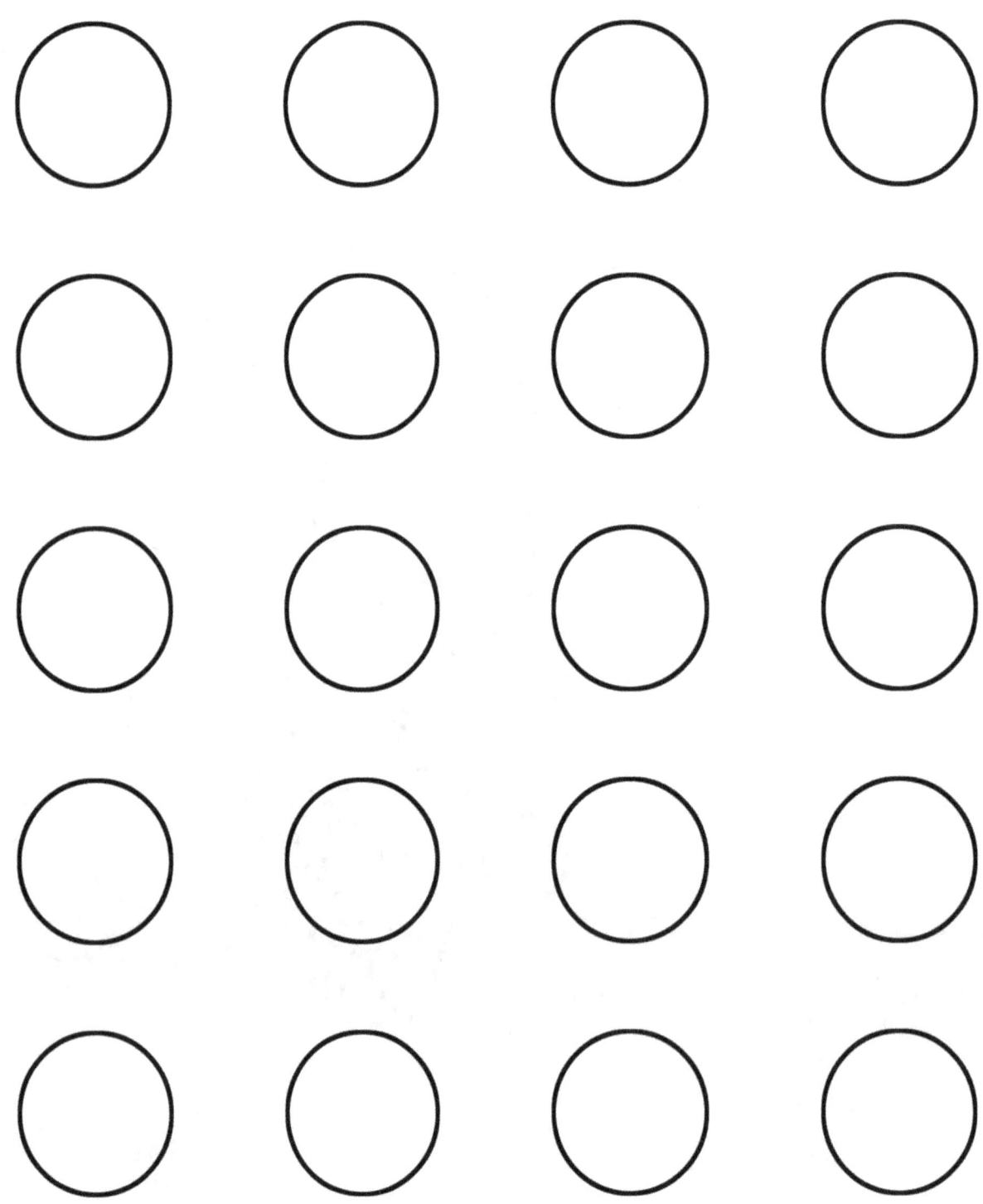

Continue to join the dots you have connected all the numbered dots. Then, color the picture!

Test Your Color

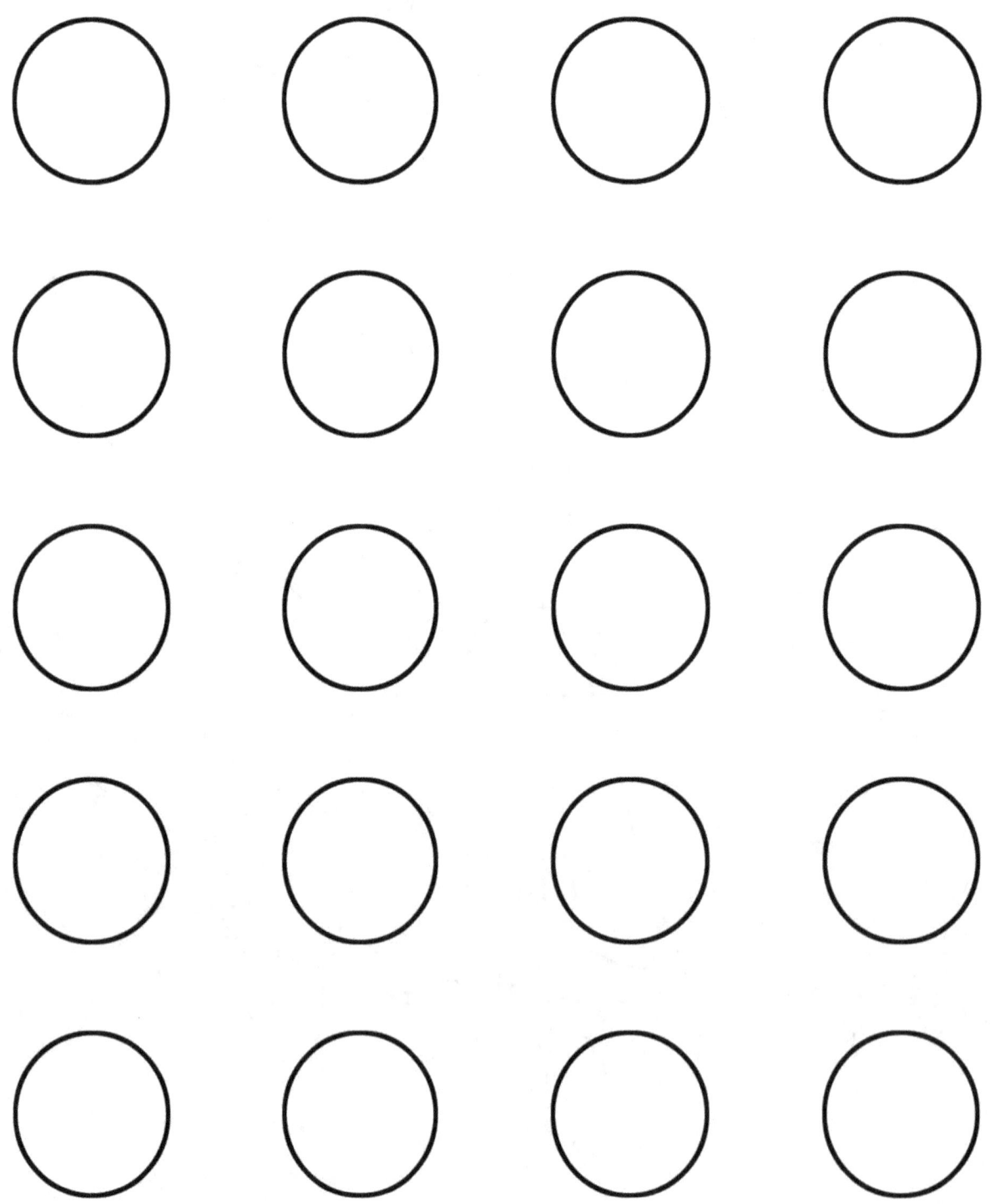

Test Your Color

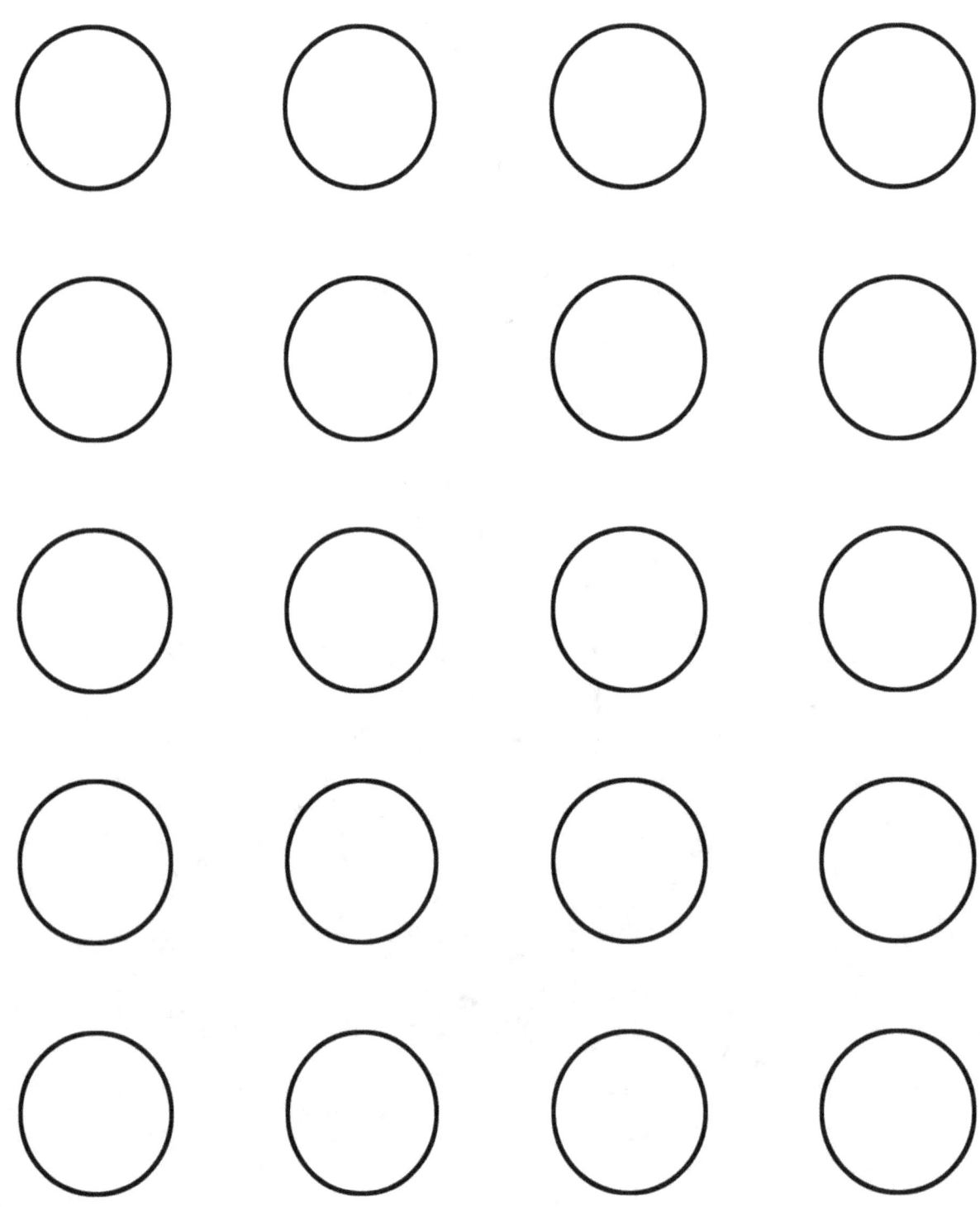

Continue to join the dots you have connected all the numbered dots. Then, color the picture!

Test Your Color

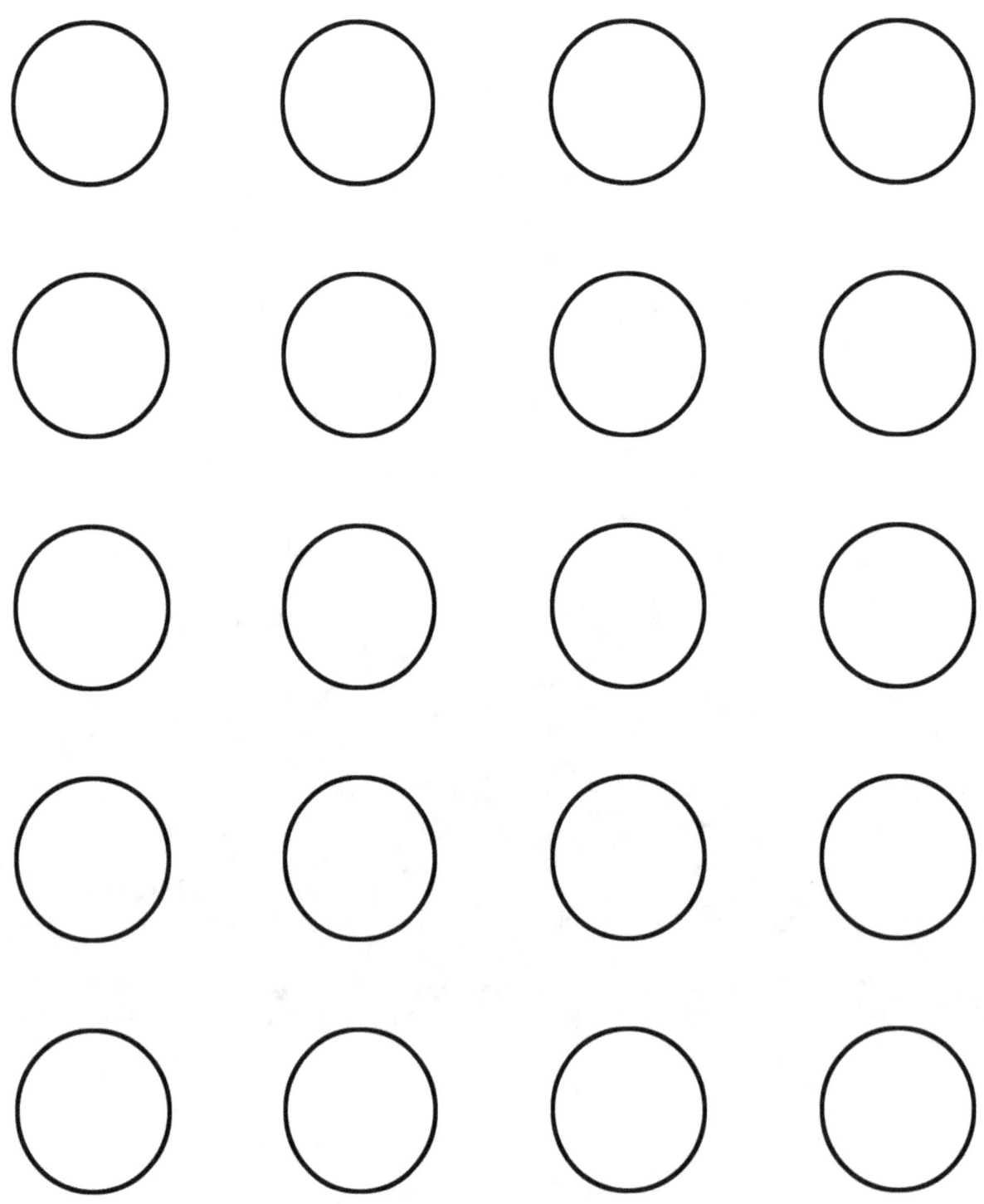

Continue to join the dots you have connected all the numbered dots. Then, color the picture!

Test Your Color

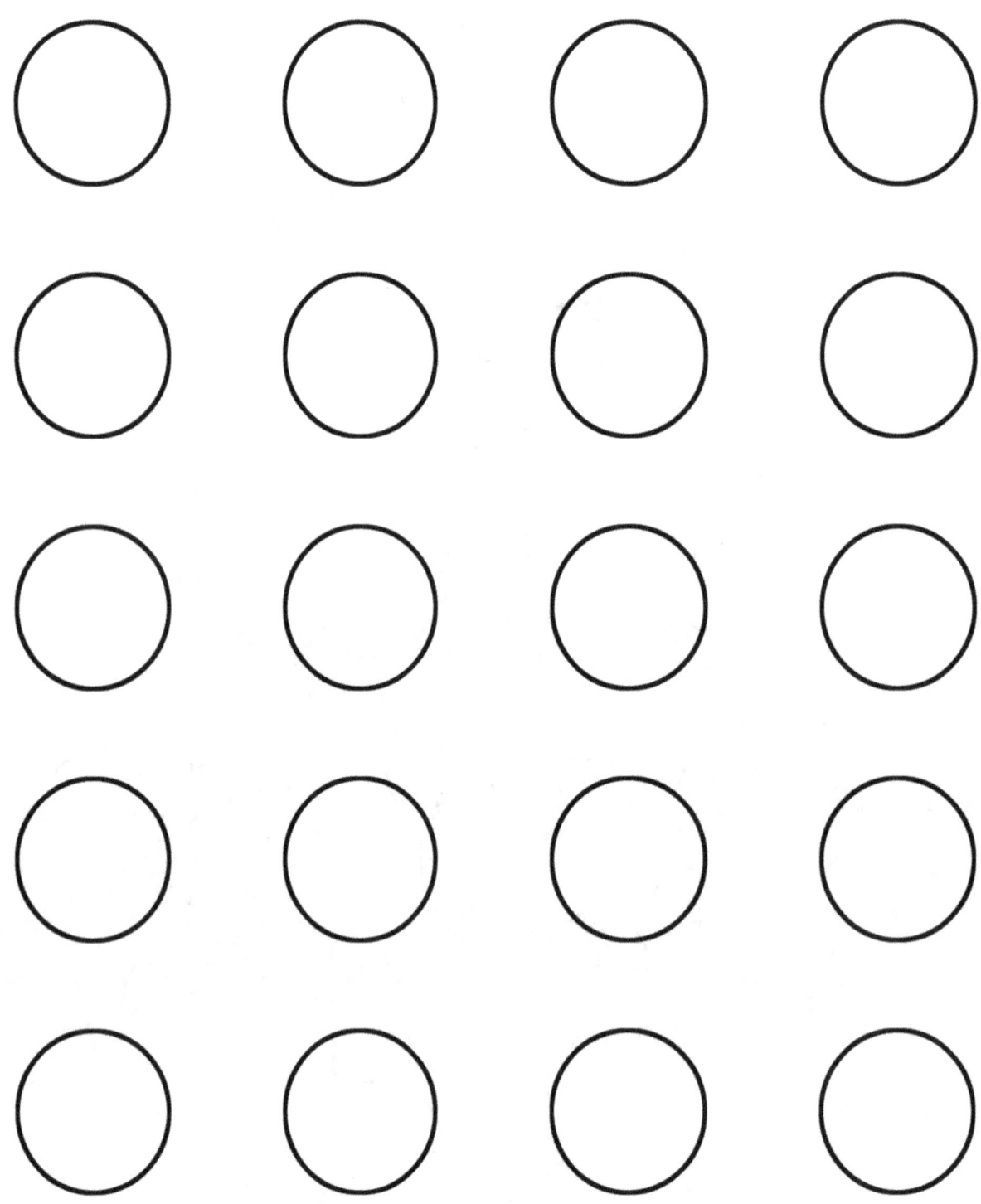

Continue to join the dots you have connected all the numbered dots. Then, color the picture!

Test Your Color

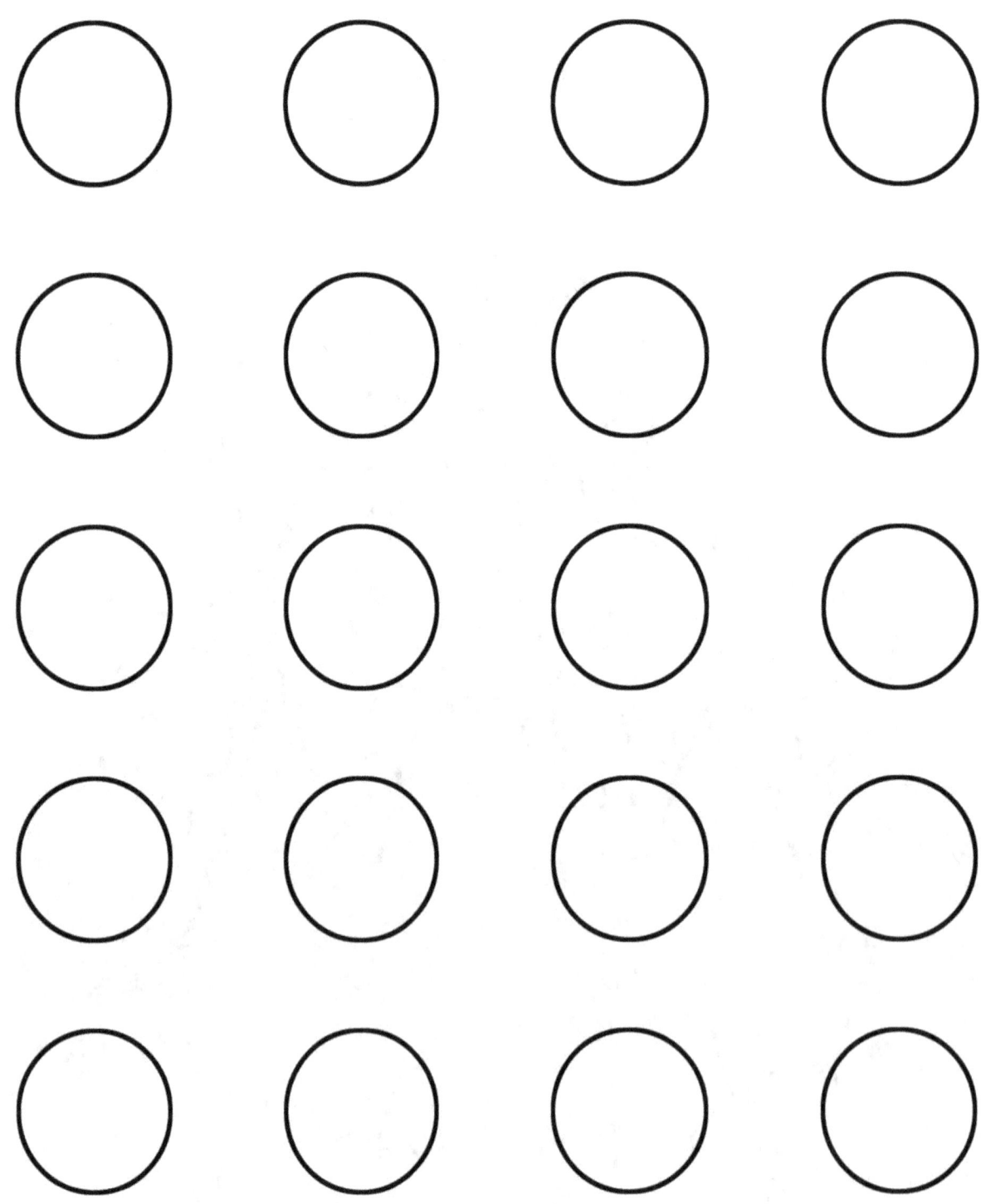

Continue to join the dots you have connected all the numbered dots. Then, color the picture!

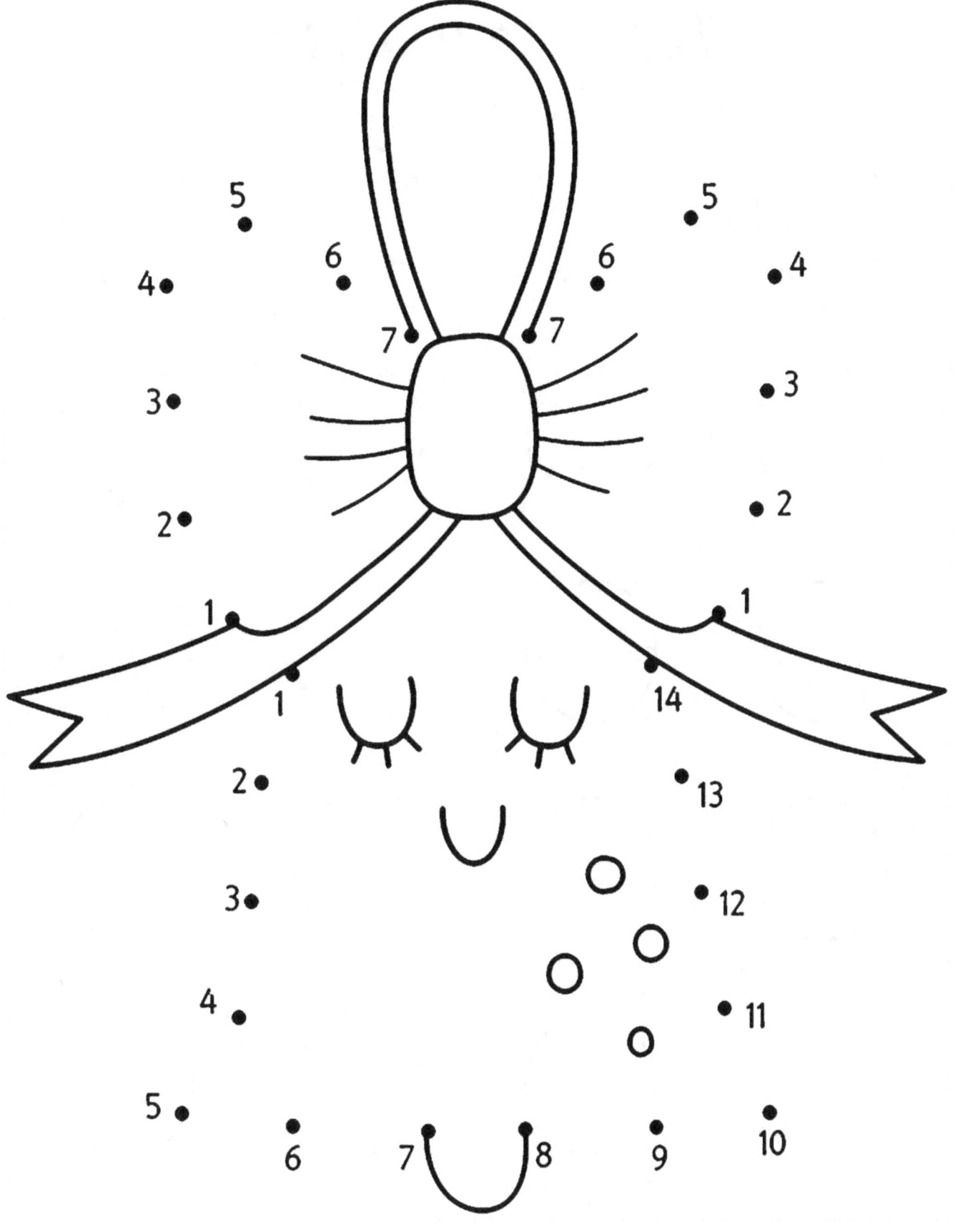

Test Your Color

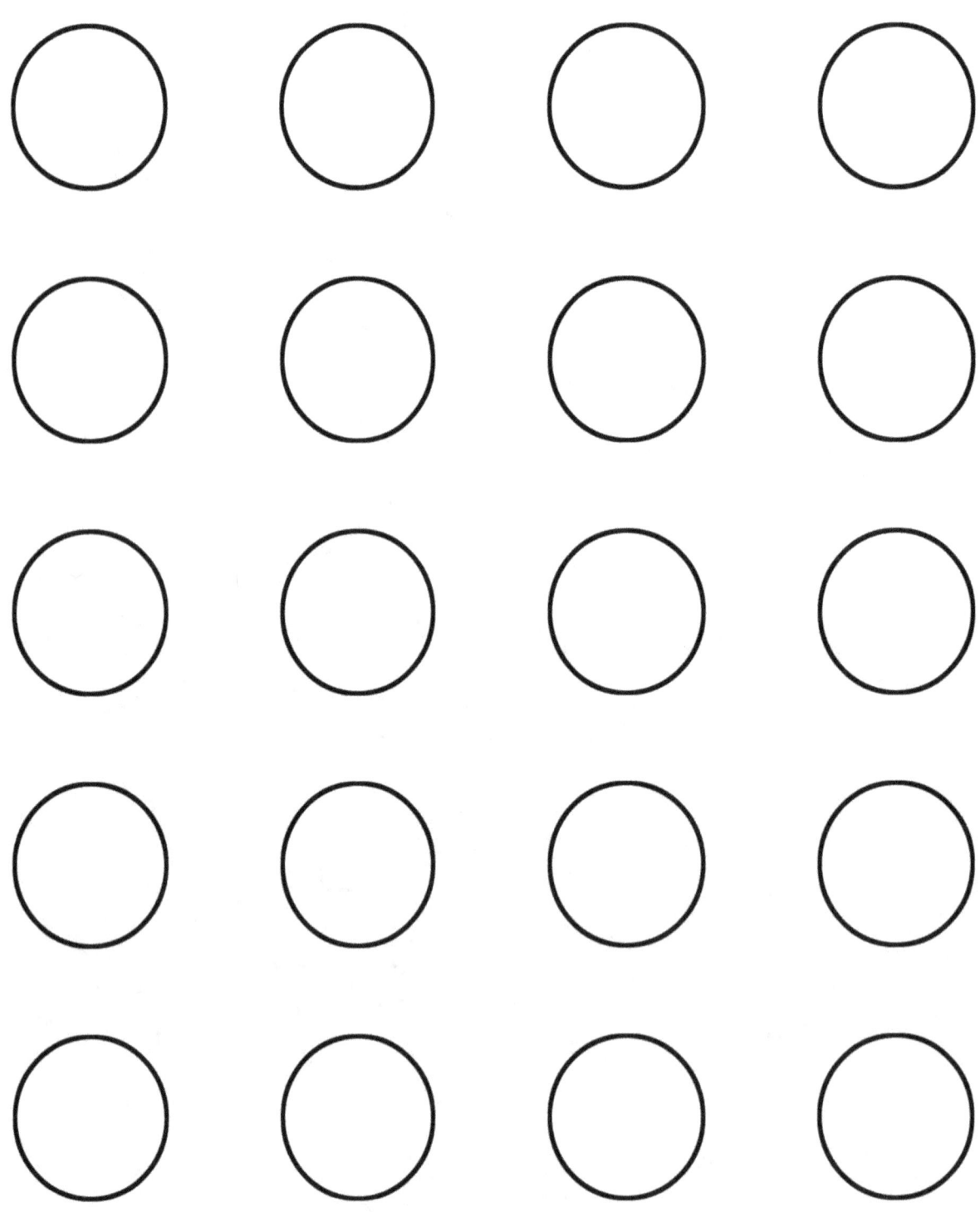

Continue to join the dots you have connected all the numbered dots. Then, color the picture!

Test Your Color

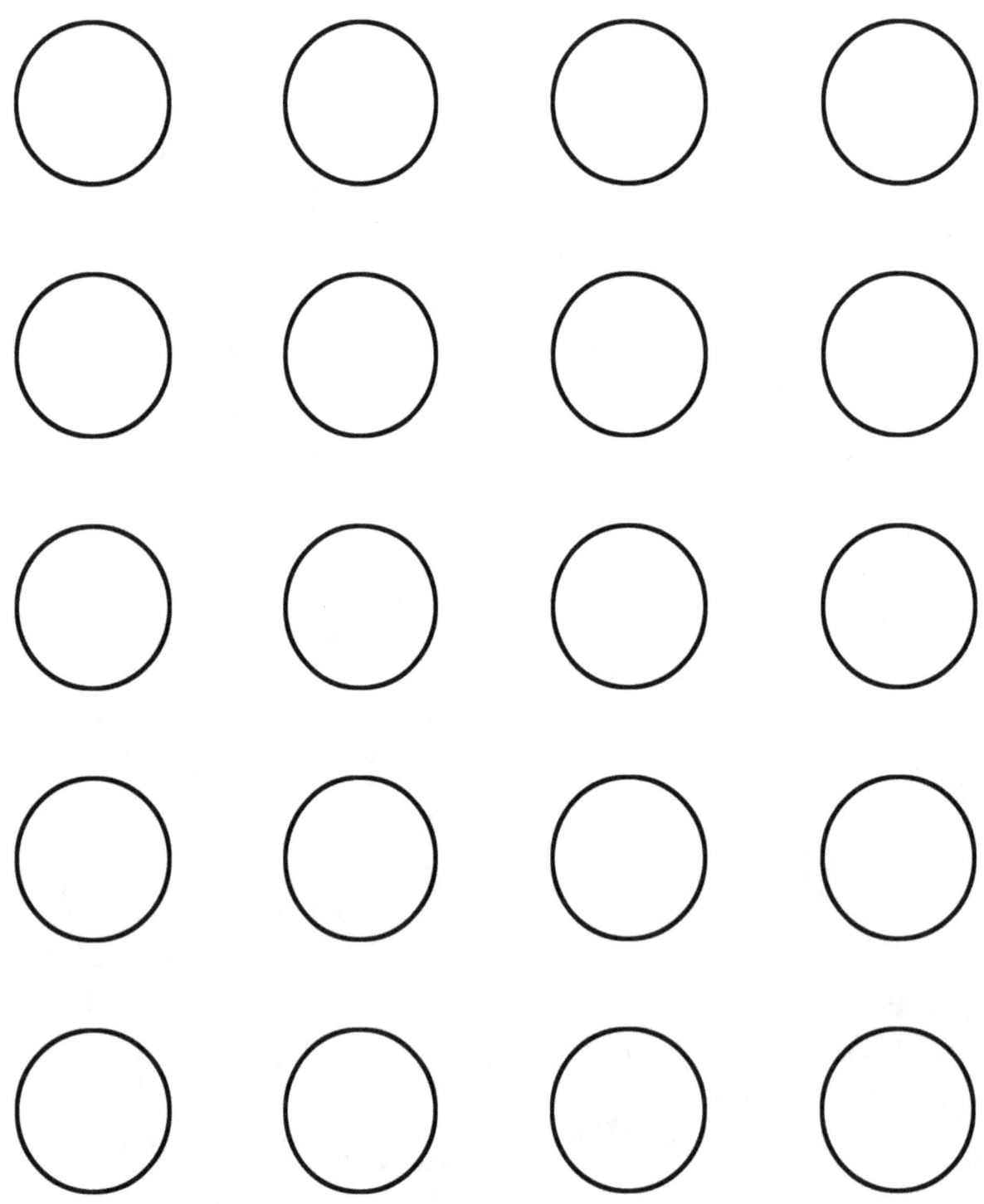

Continue to join the dots you have connected all the numbered dots. Then, color the picture!

Test Your Color

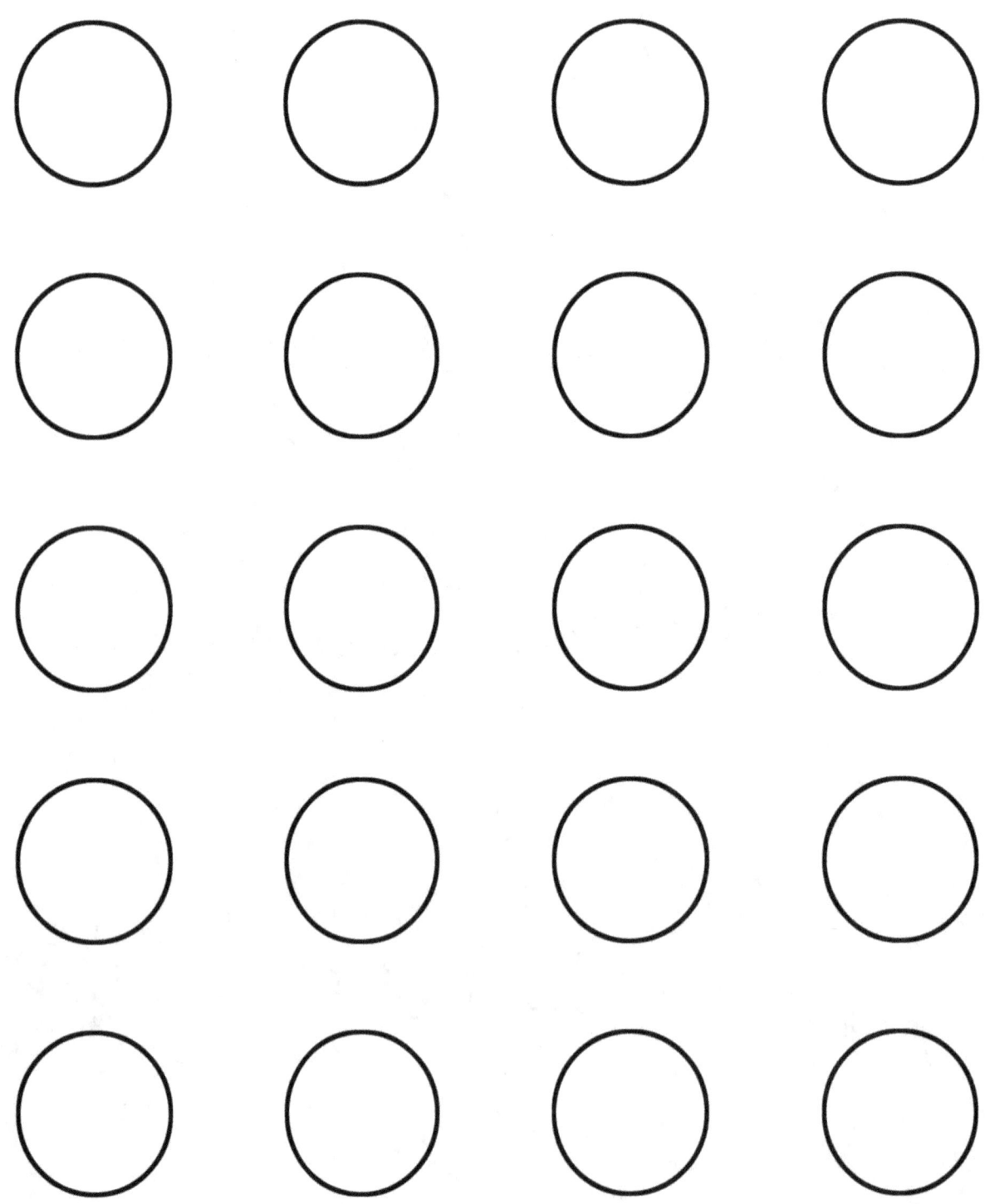

Continue to join the dots you have connected all the numbered dots. Then, color the picture!

Test Your Color

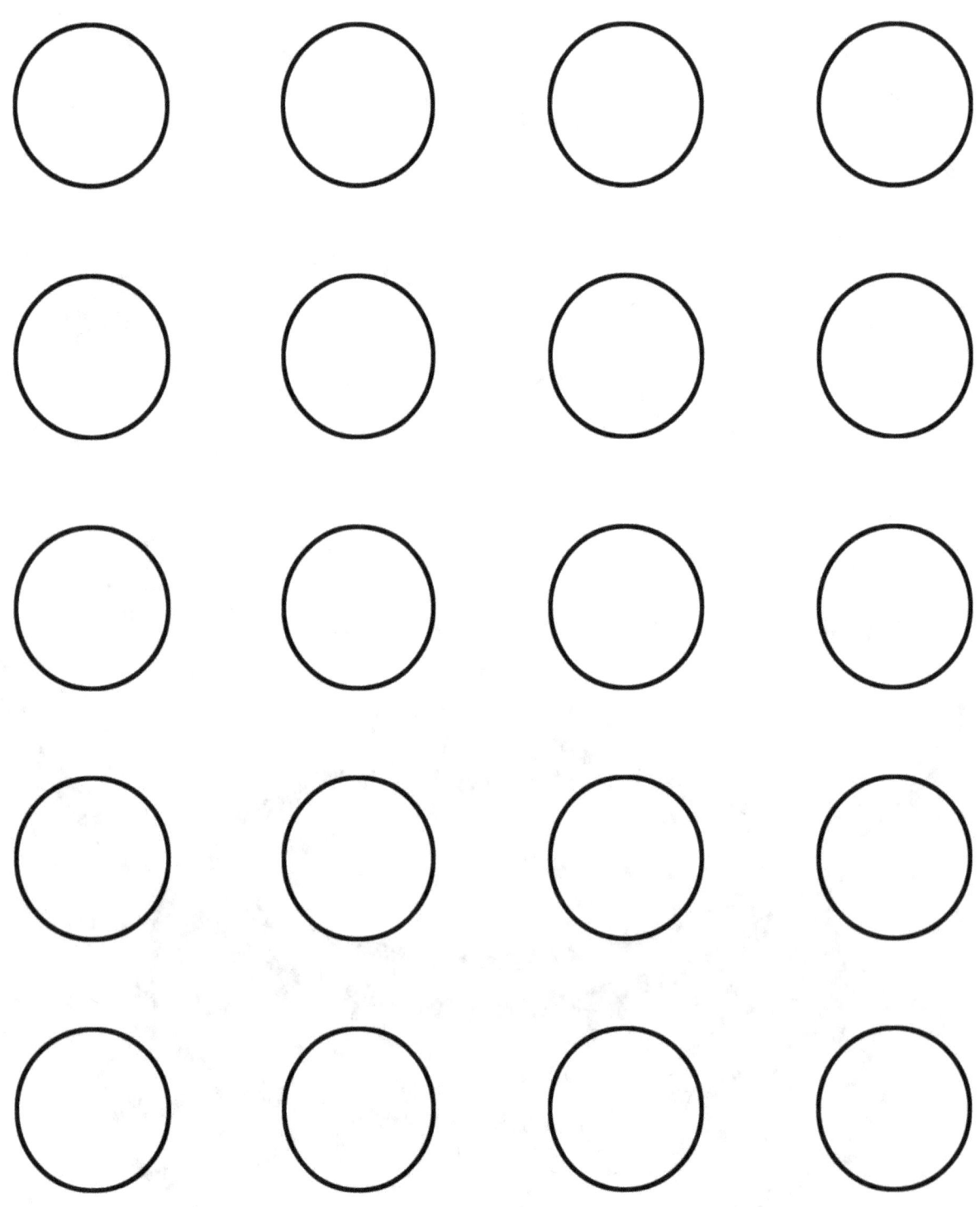

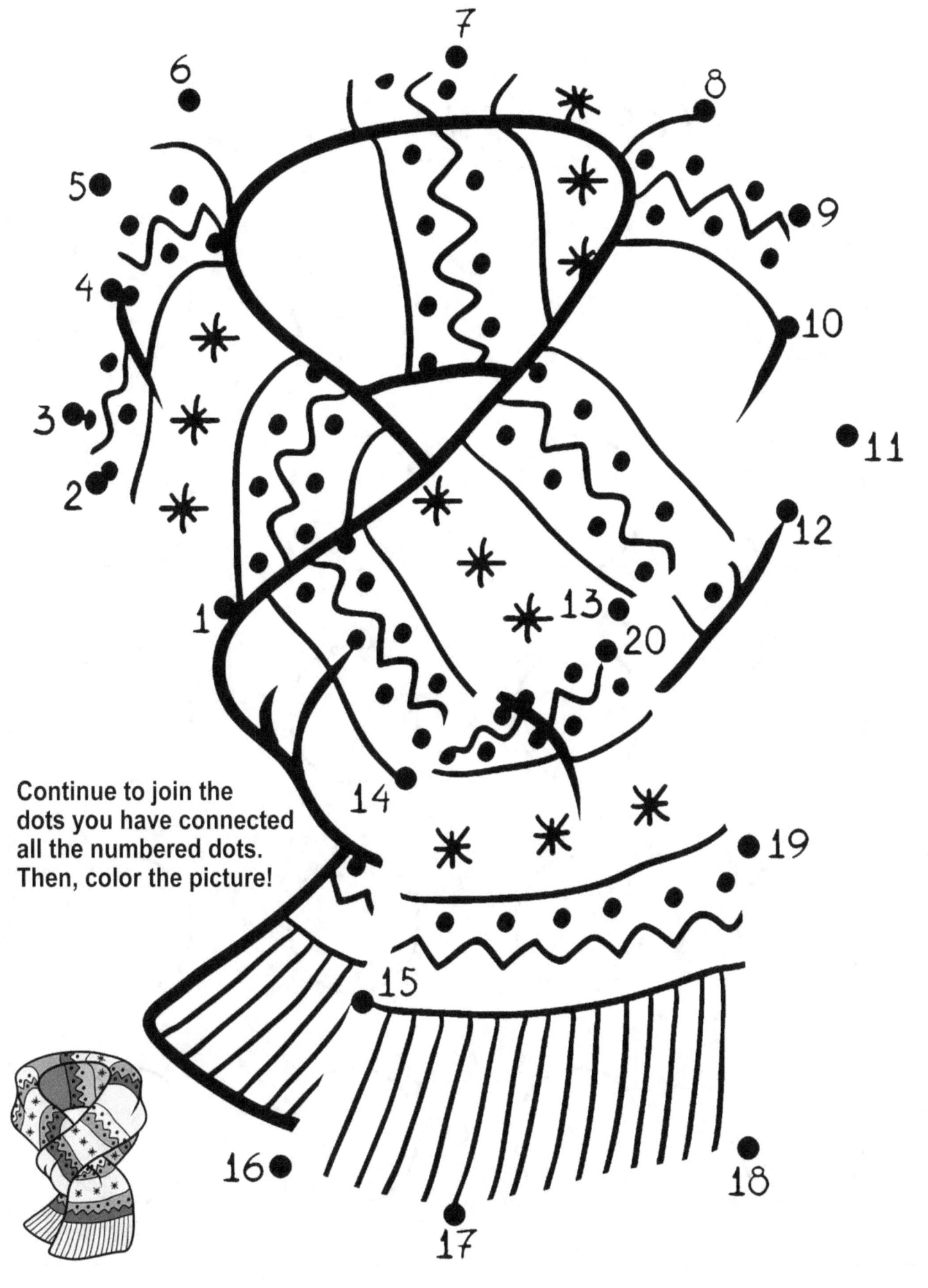

Continue to join the dots you have connected all the numbered dots. Then, color the picture!

Test Your Color

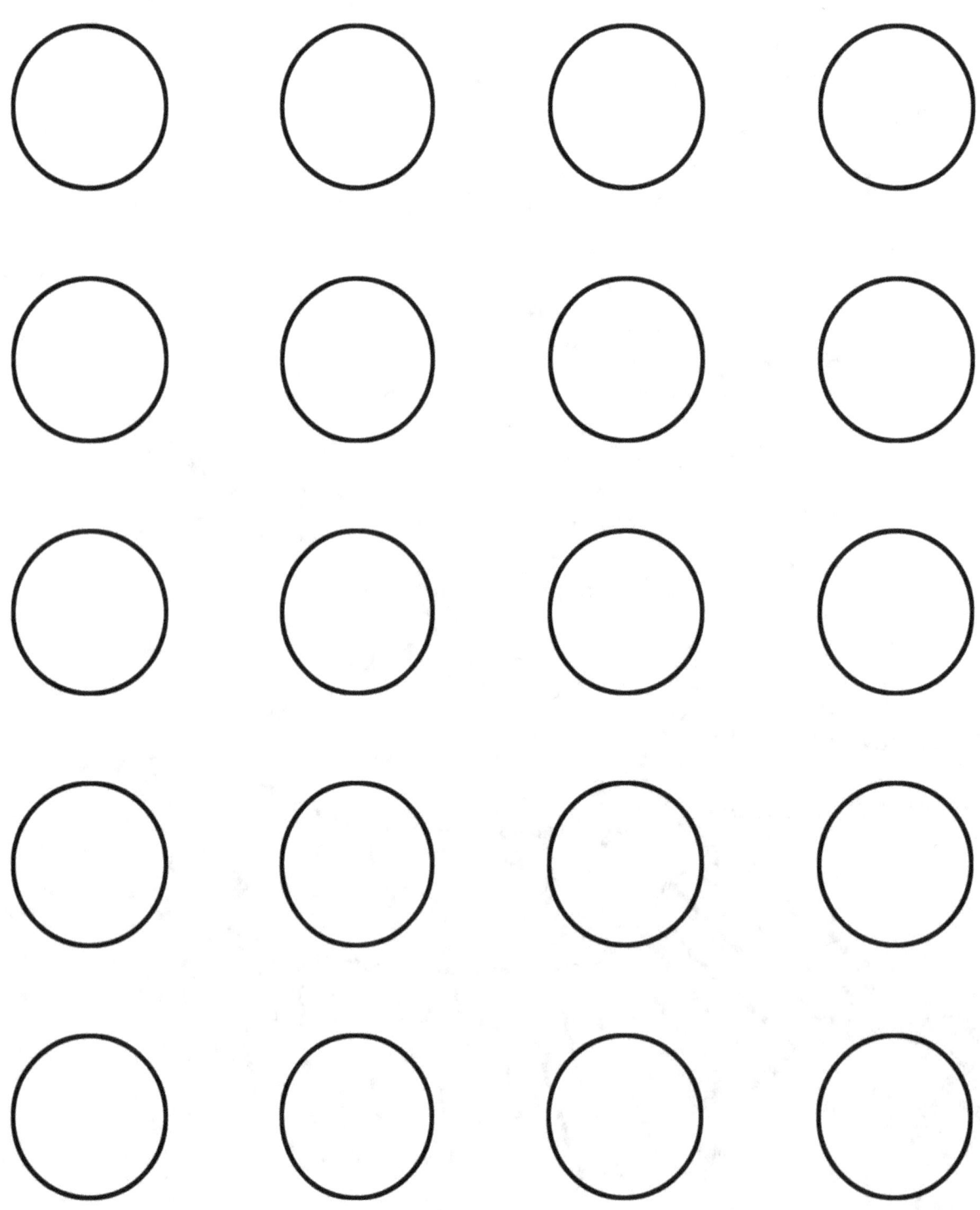

Test Your Color

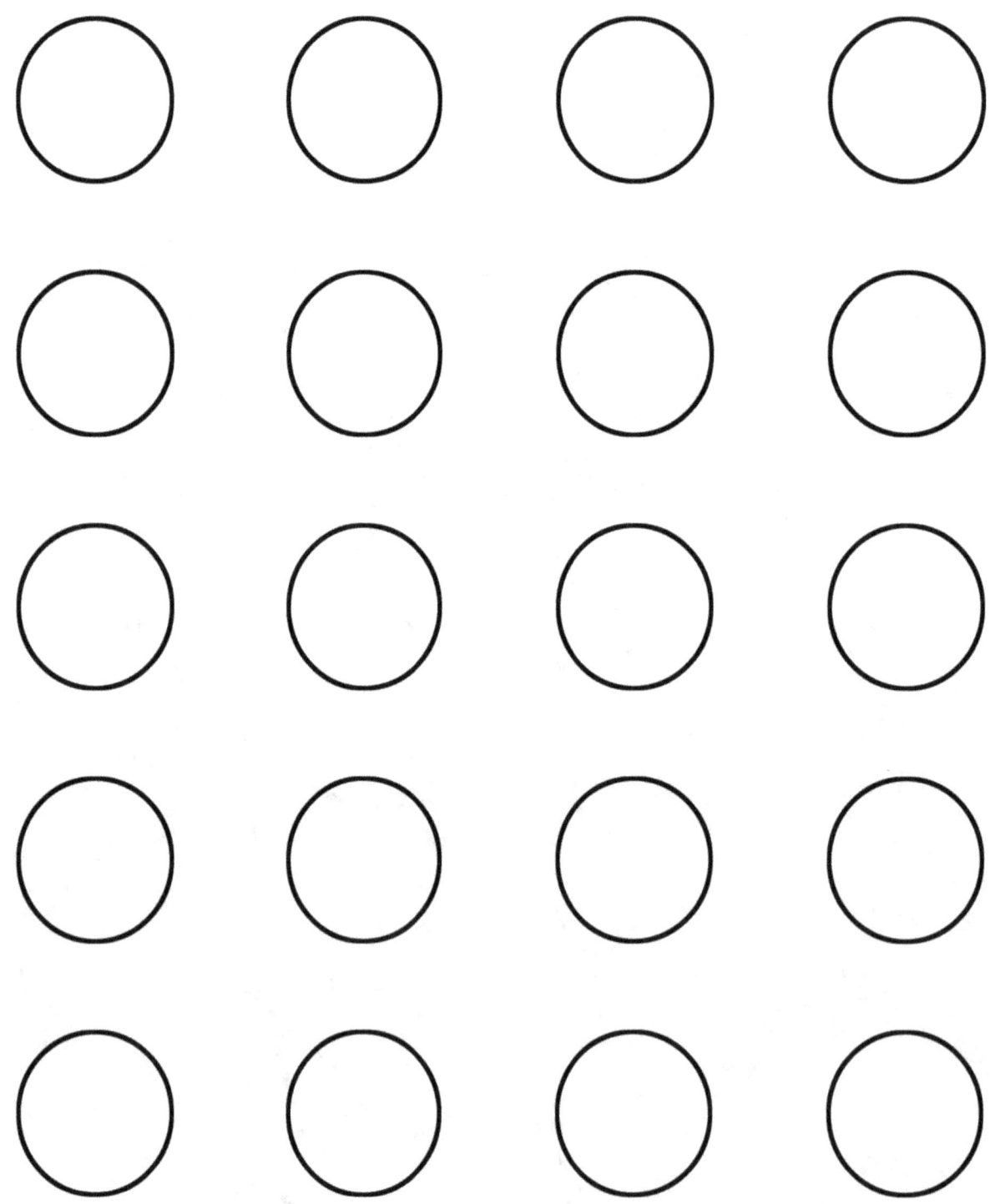

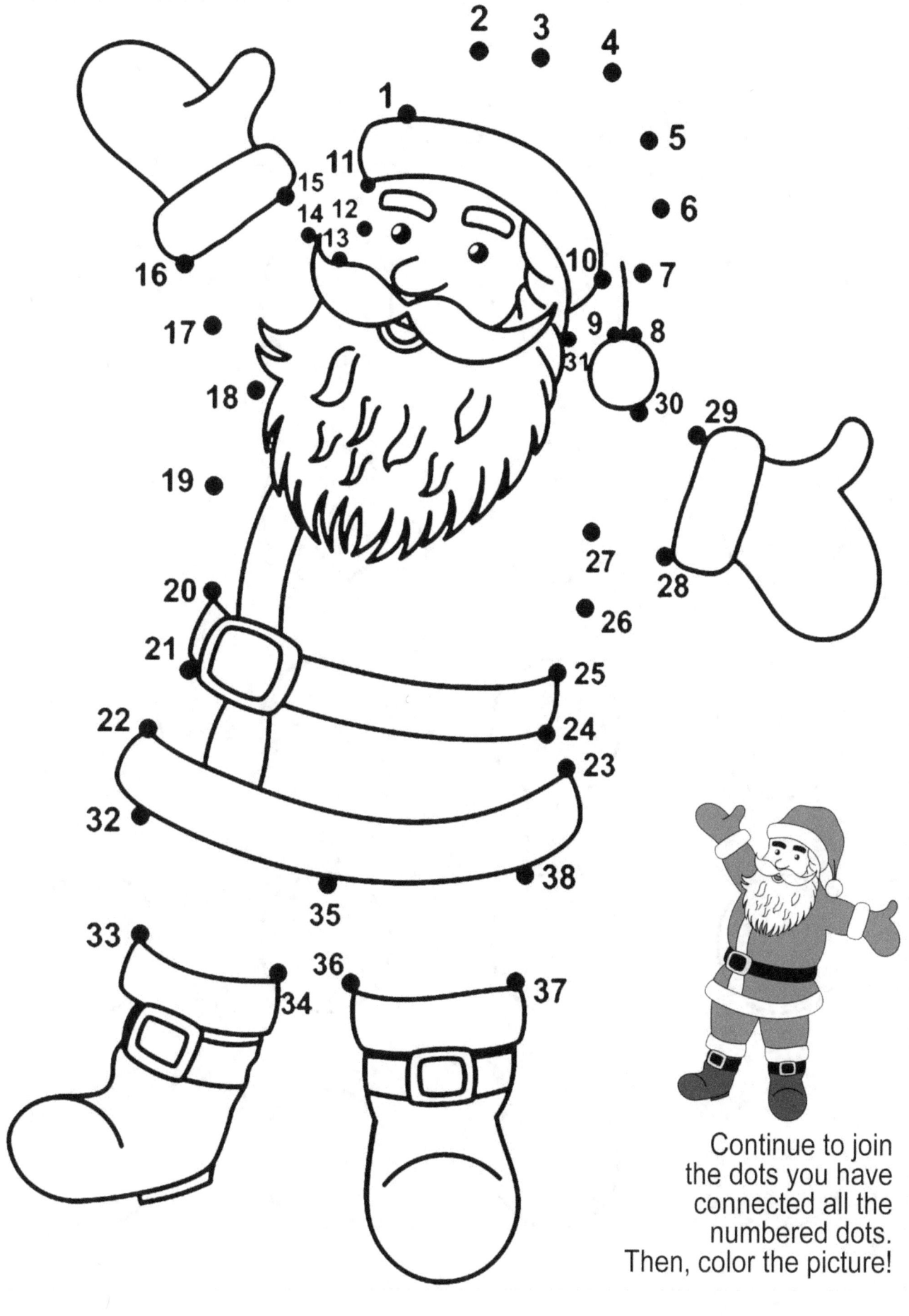

Continue to join the dots you have connected all the numbered dots. Then, color the picture!

Test Your Color

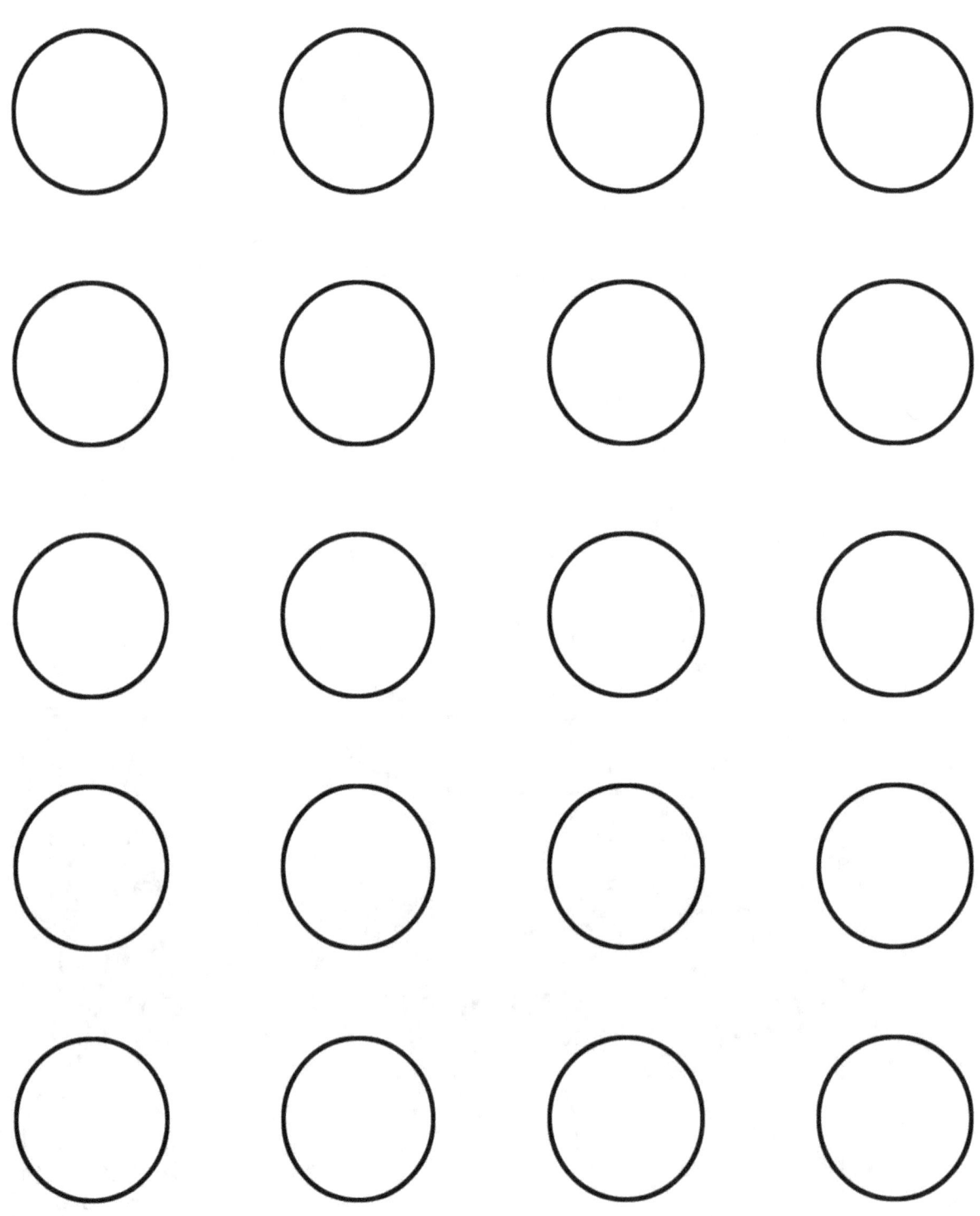

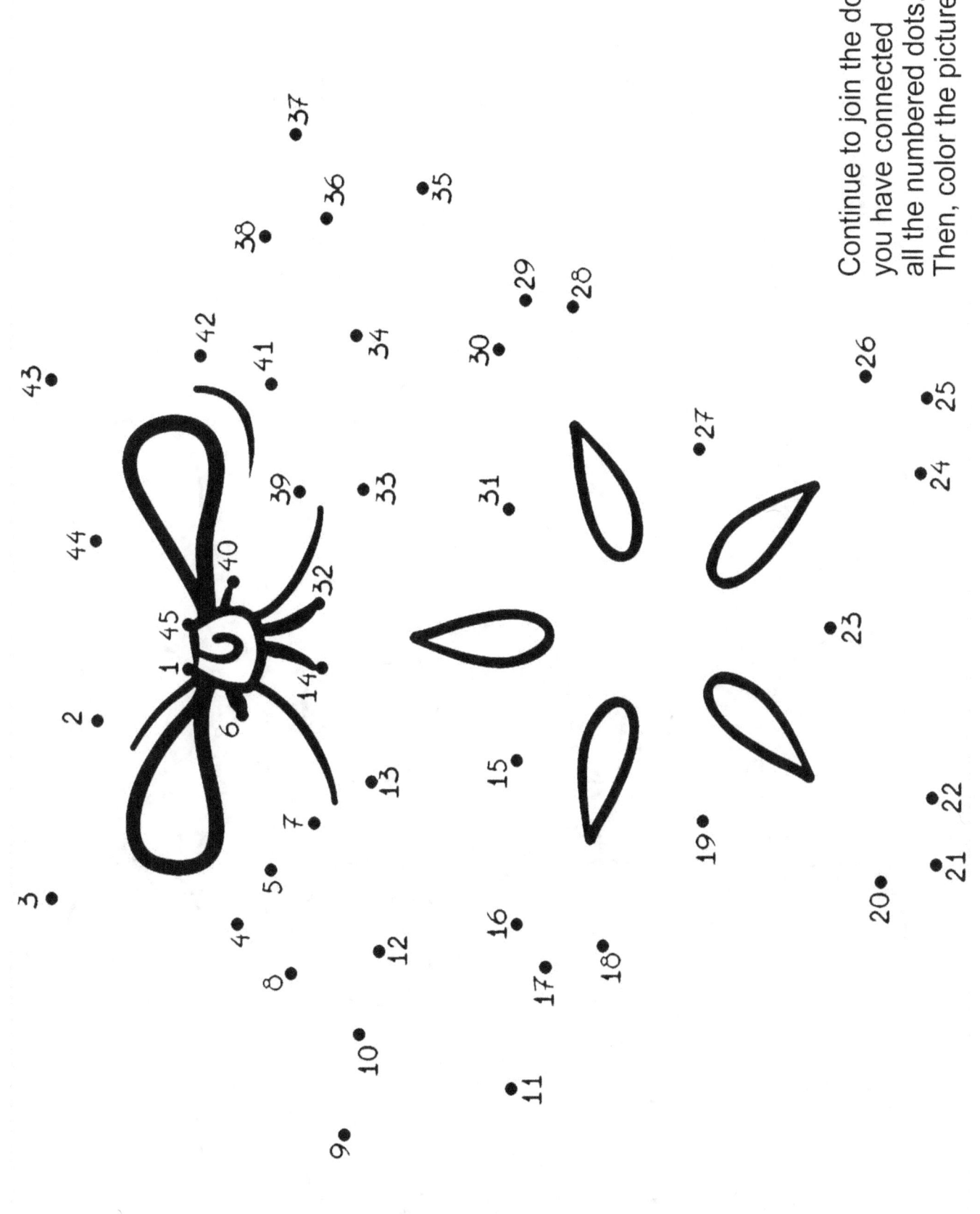

Continue to join the dots you have connected all the numbered dots. Then, color the picture!

Test Your Color

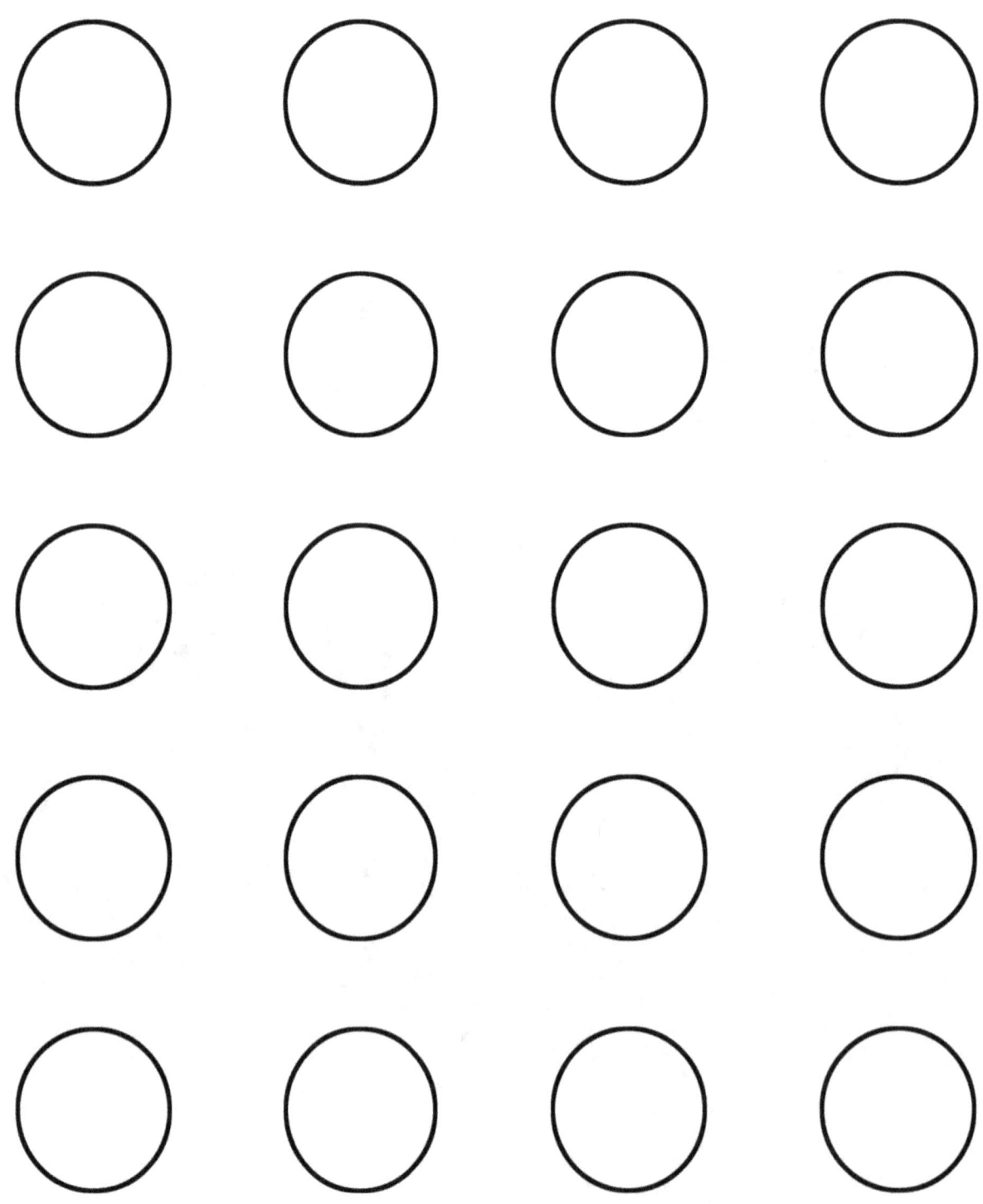

Continue to join the dots you have connected all the numbered dots. Then, color the picture!

Test Your Color

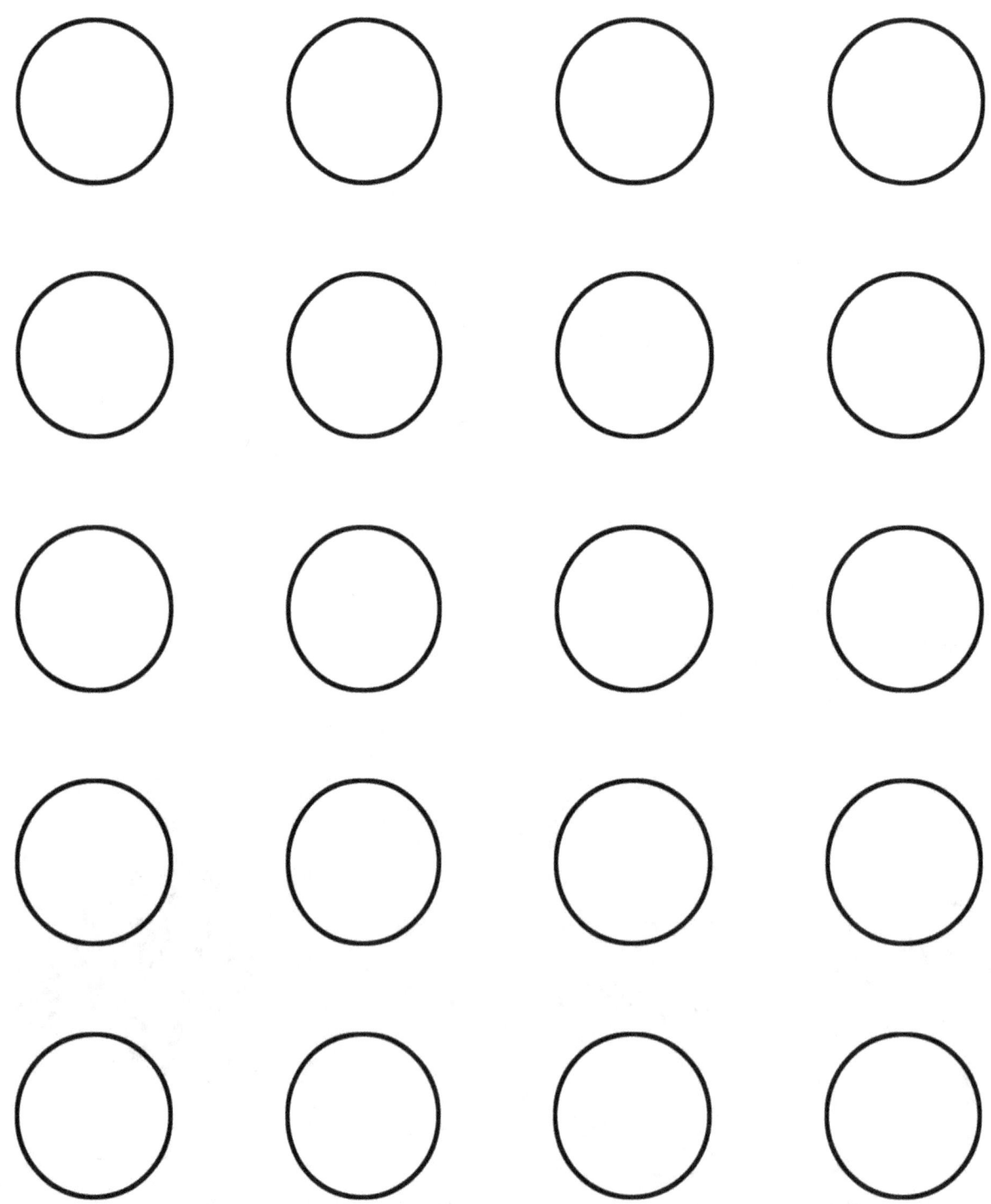

Continue to join the dots you have connected all the numbered dots. Then, color the picture!

Test Your Color

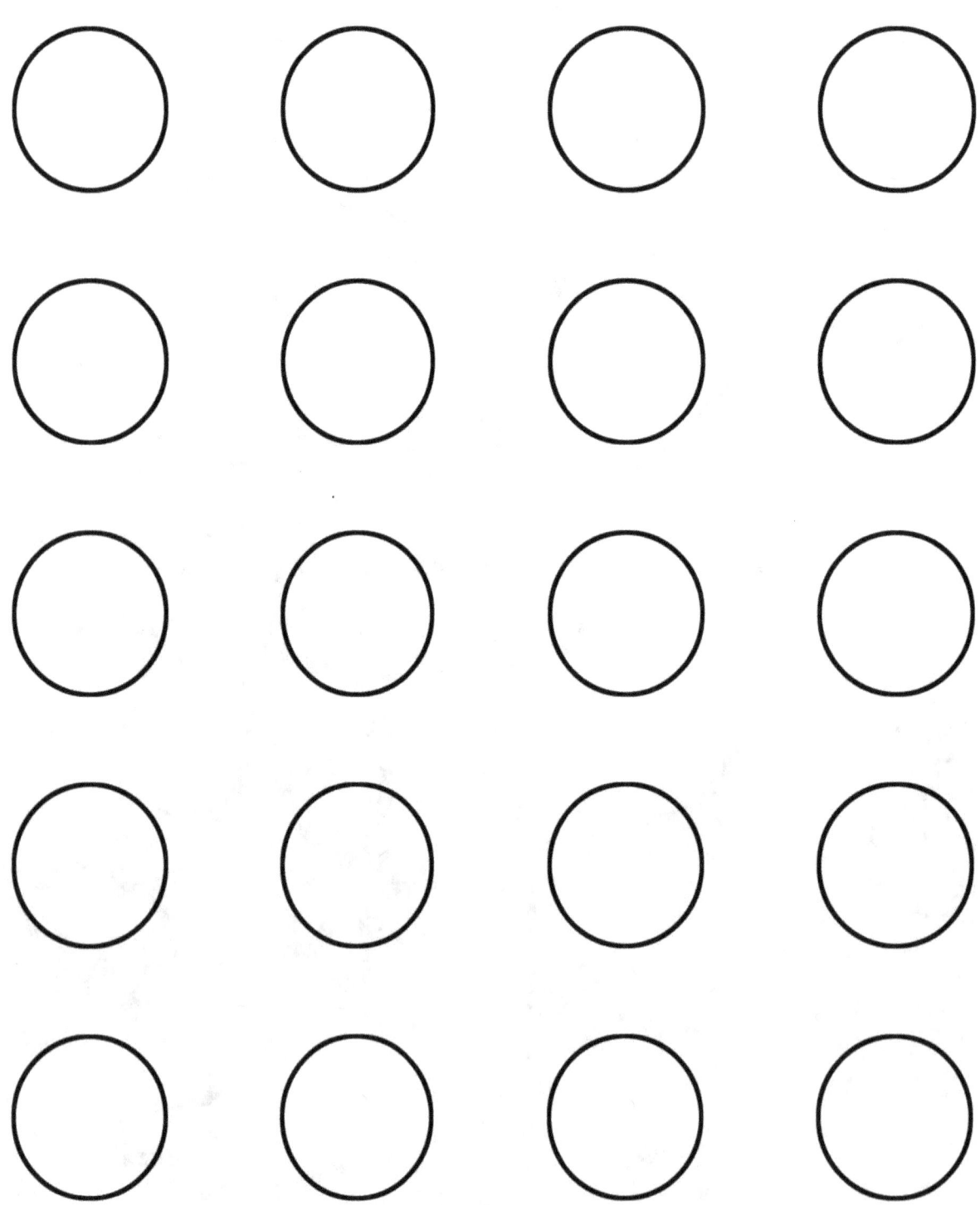

Continue to join the dots you have connected all the numbered dots. Then, color the picture!

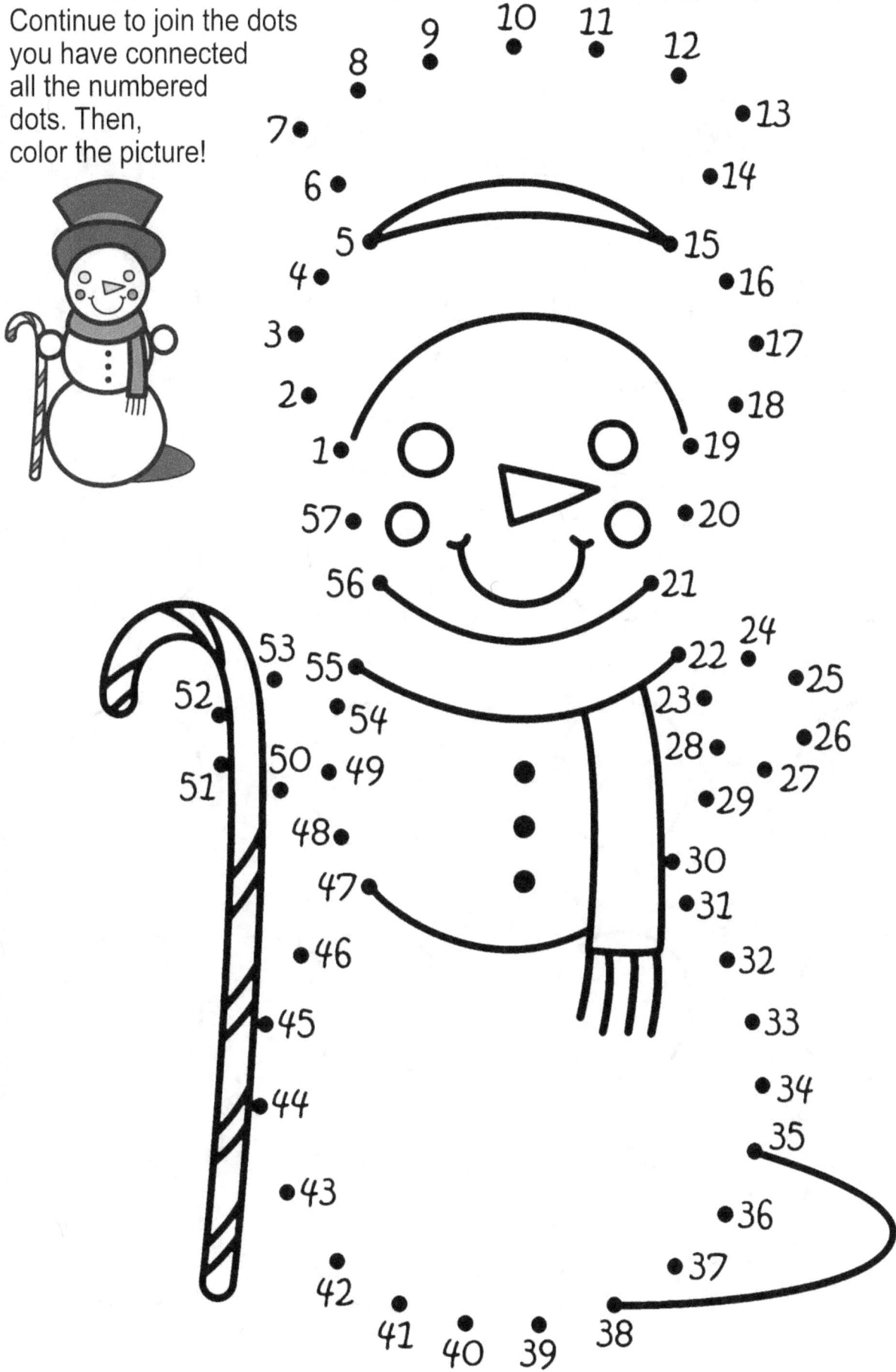

Test Your Color

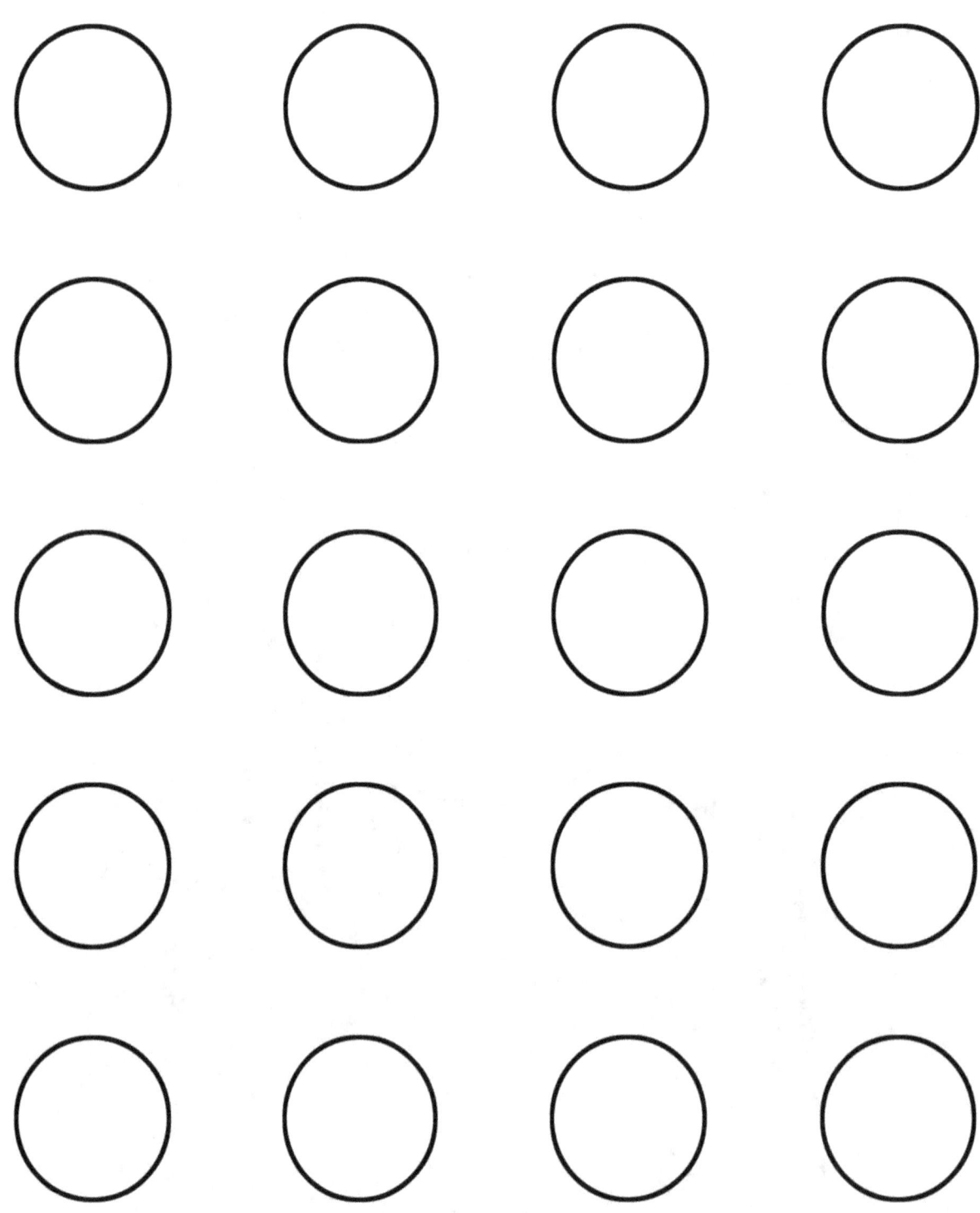

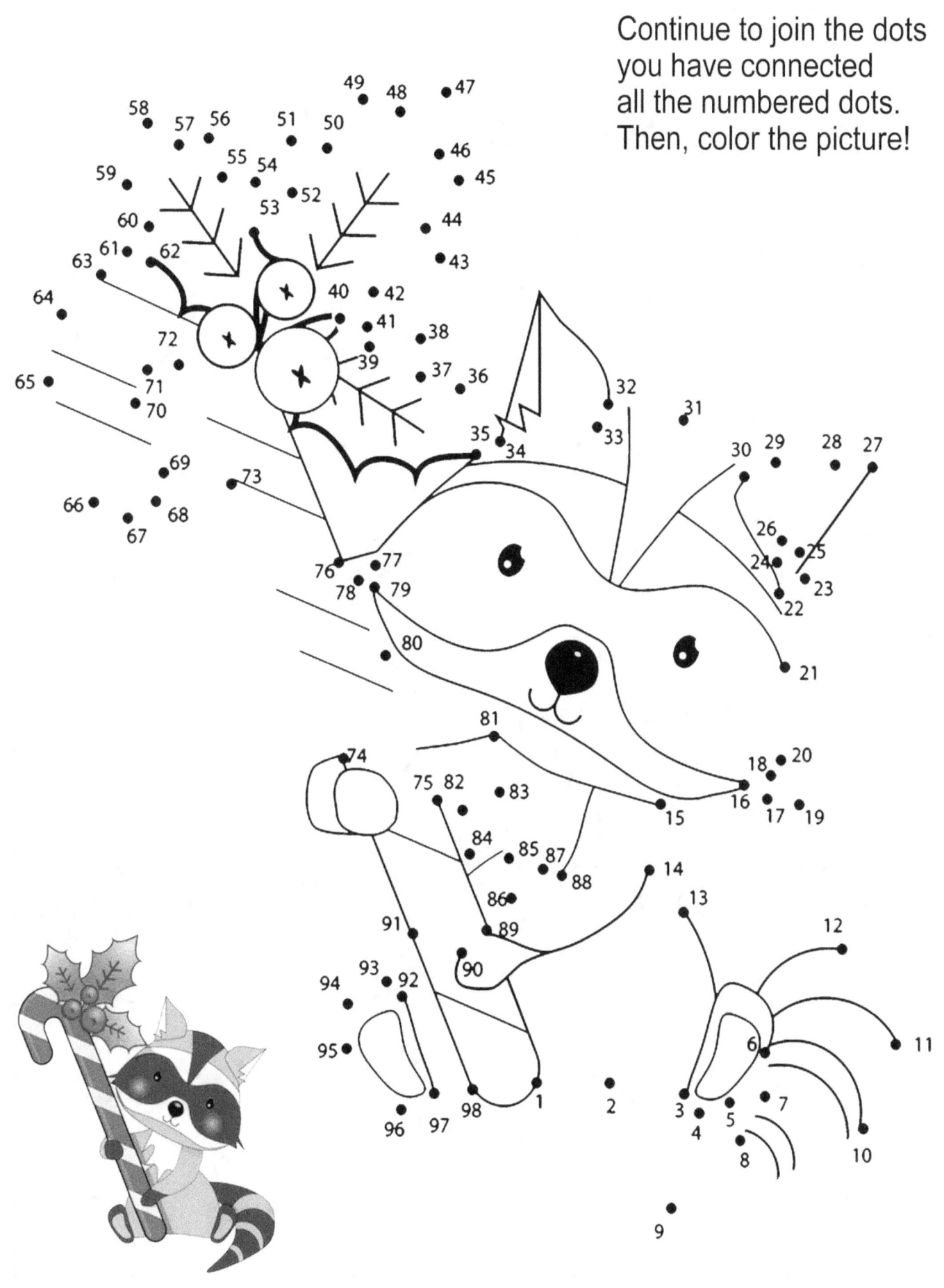

Test Your Color

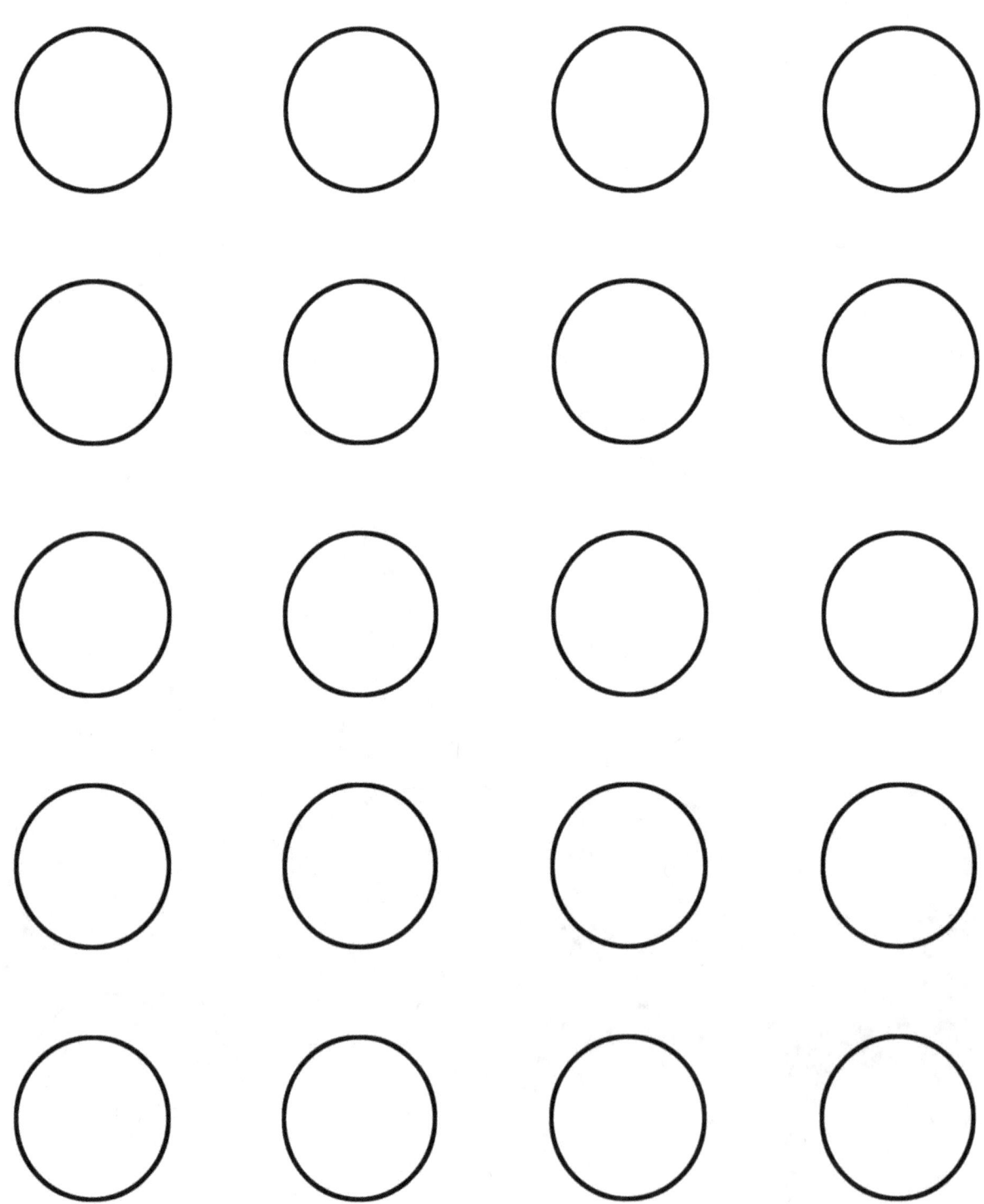

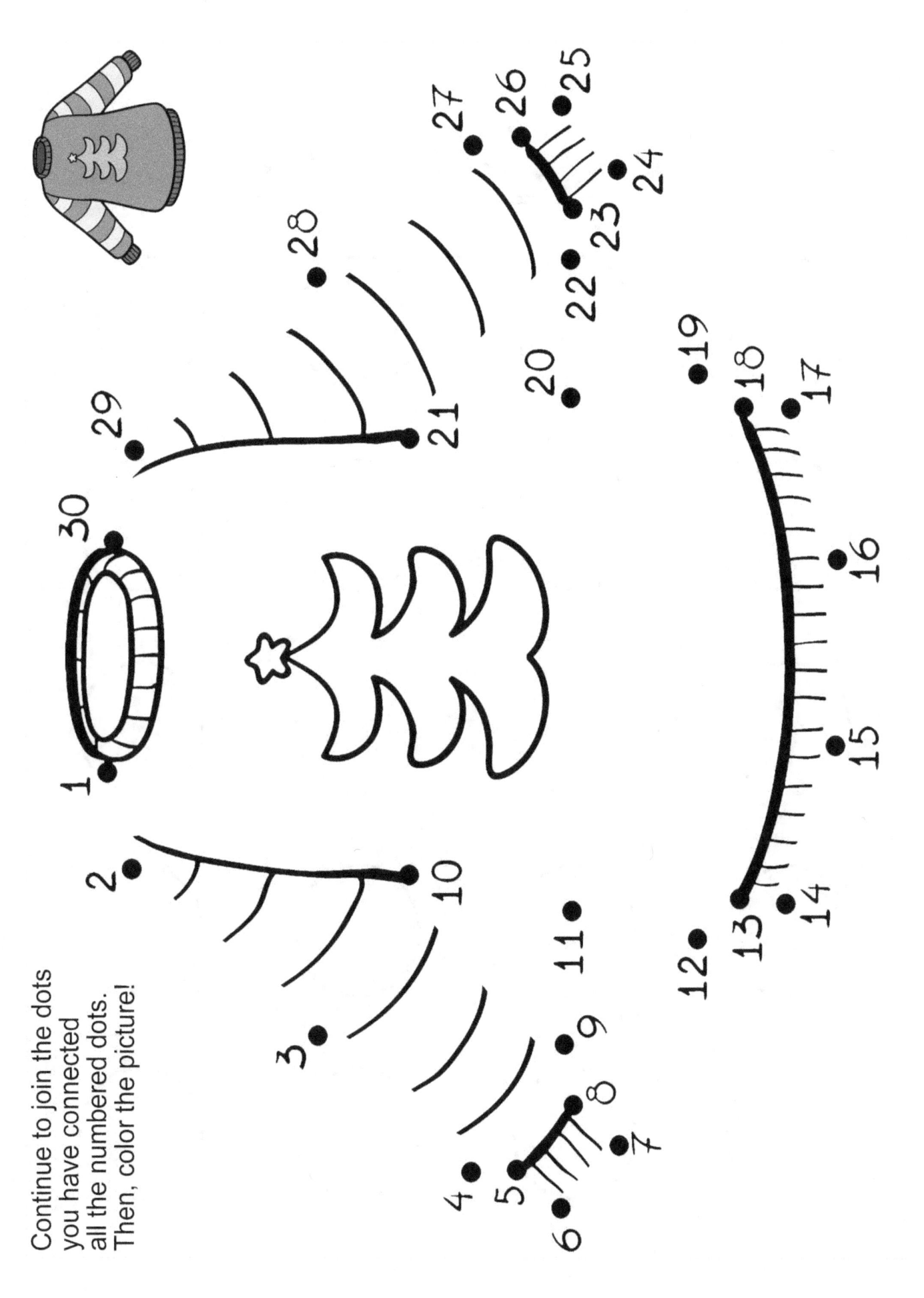

Test Your Color

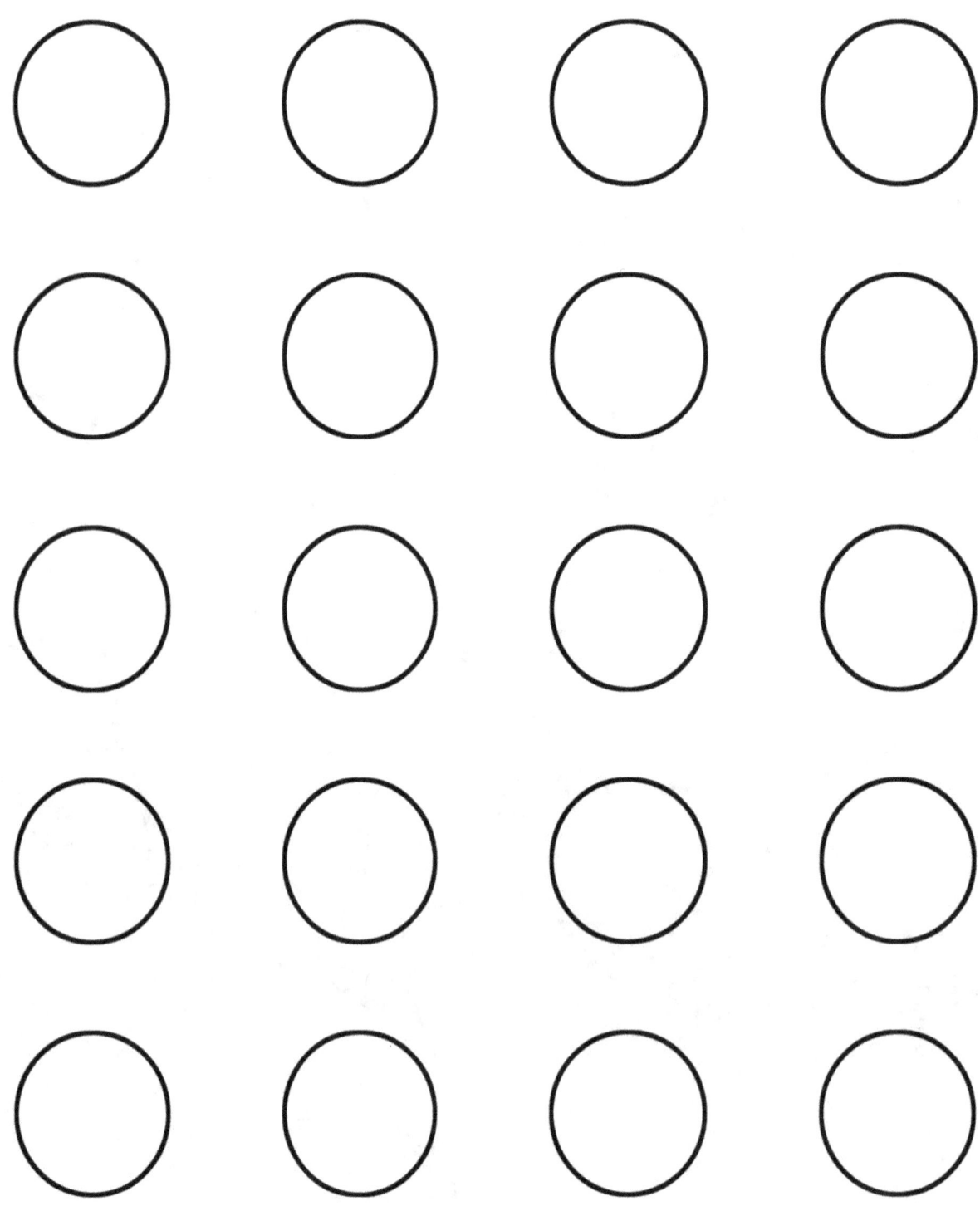

Continue to join the dots you have connected all the numbered dots. Then, color the picture!

Test Your Color

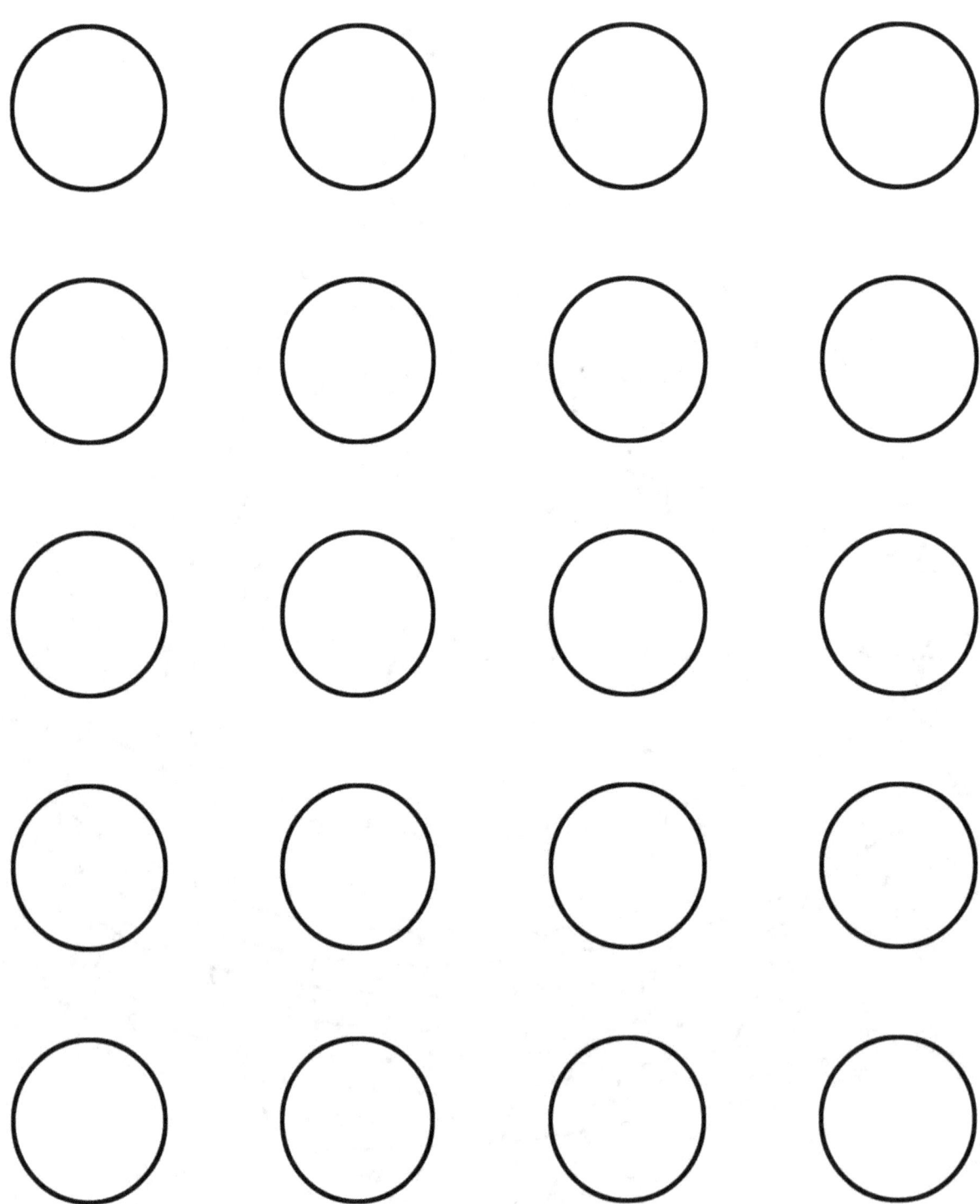

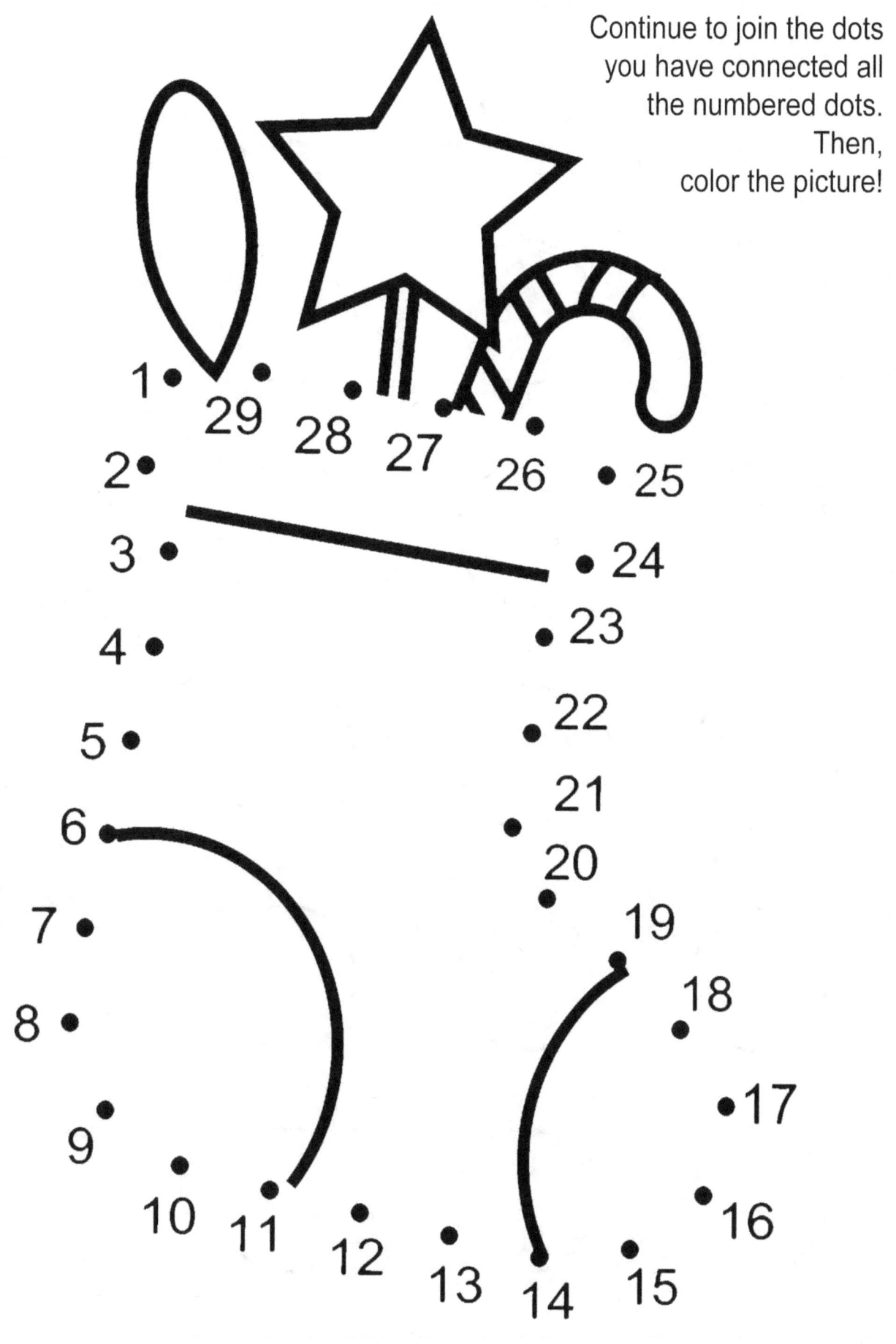

Test Your Color

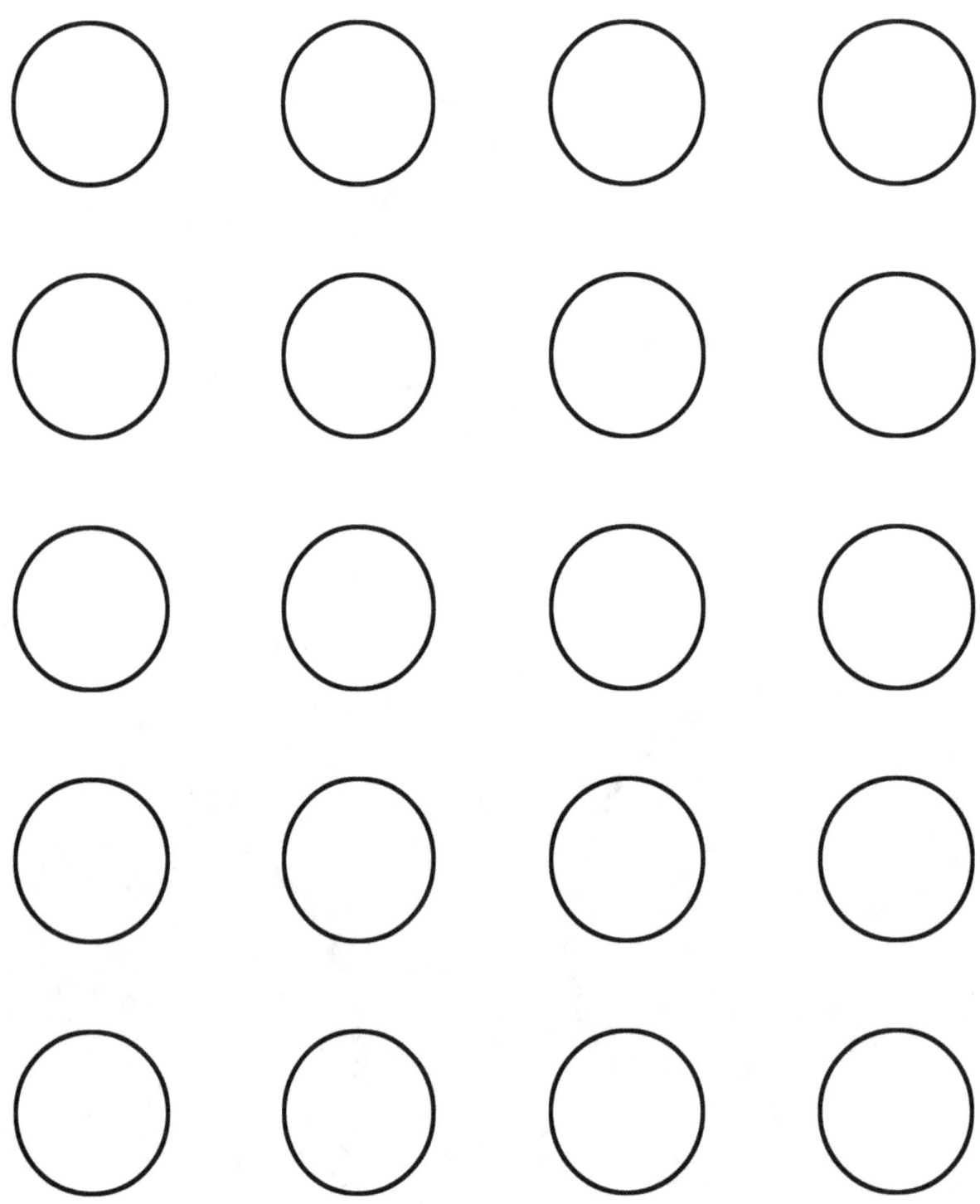

Continue to join the dots you have connected all the numbered dots. Then, color the picture!

Test Your Color

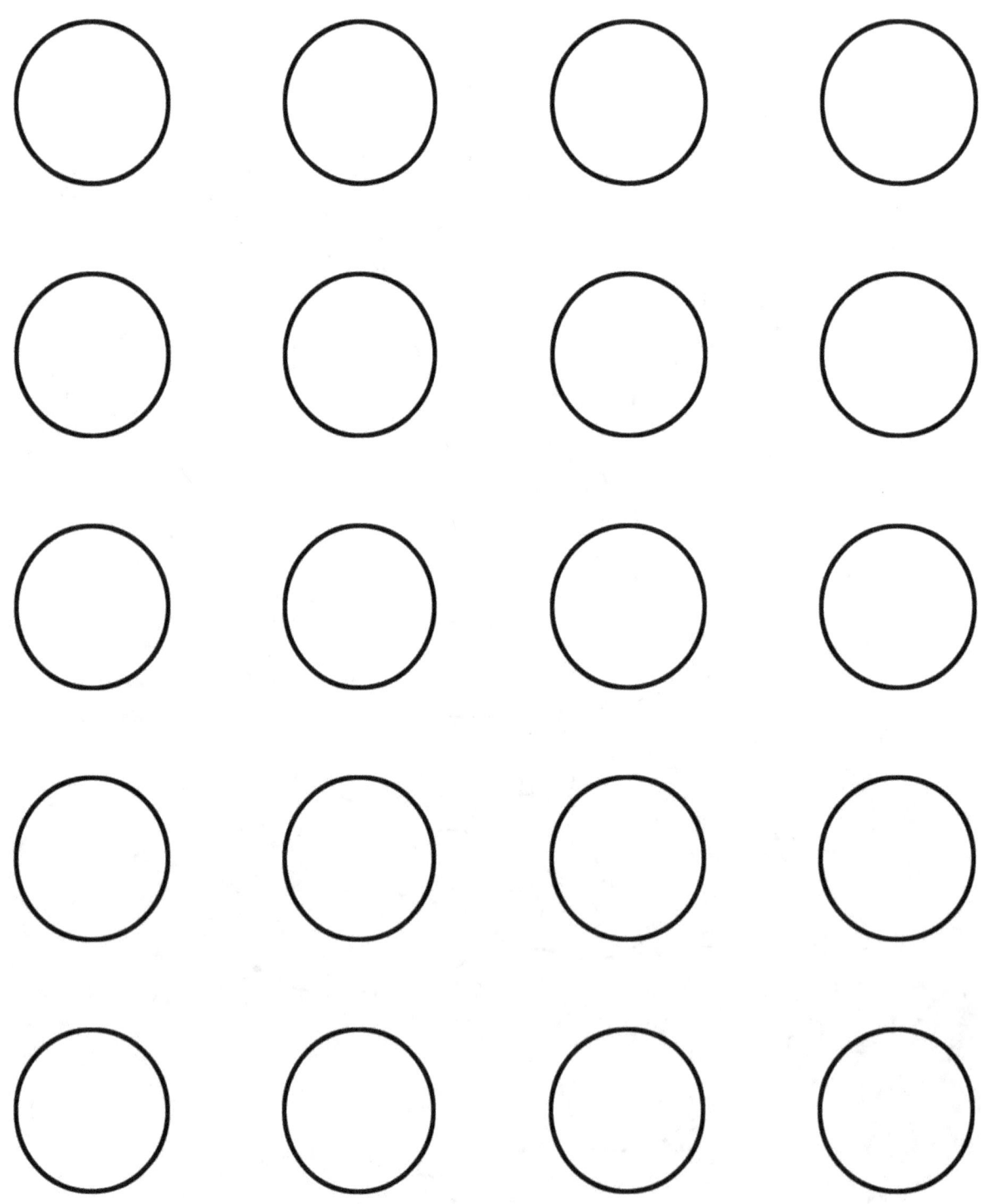

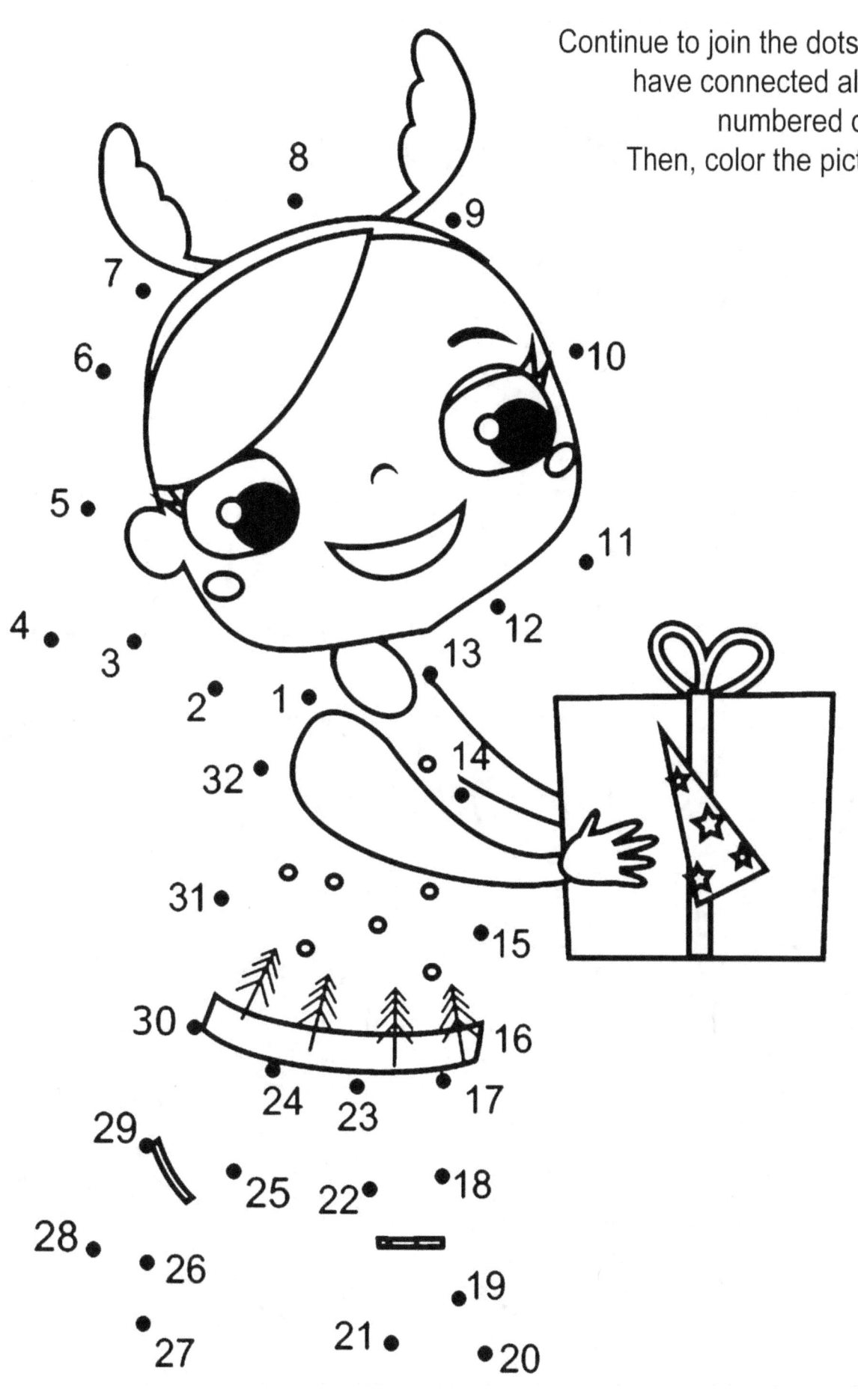

Continue to join the dots you have connected all the numbered dots. Then, color the picture!

Test Your Color

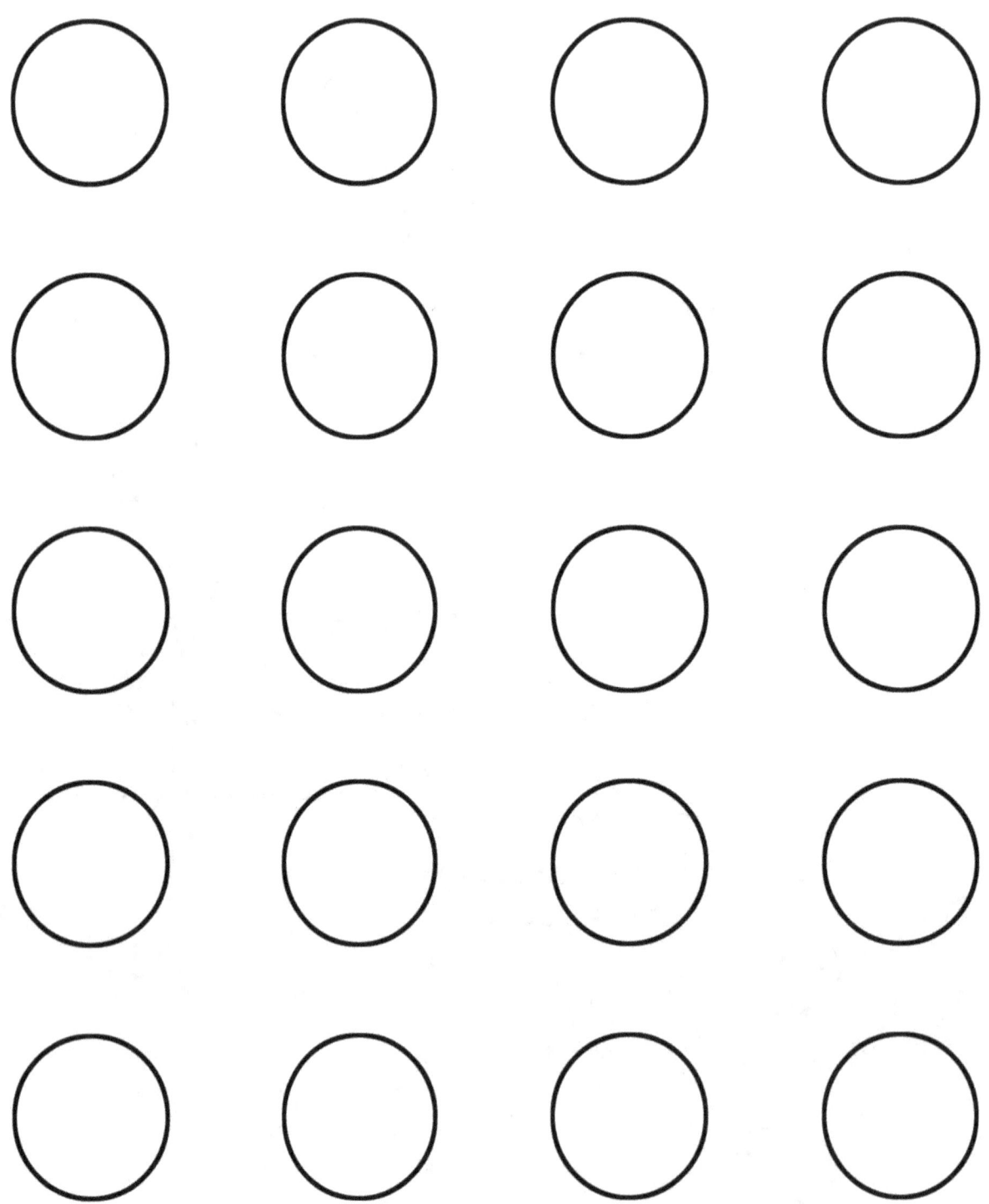

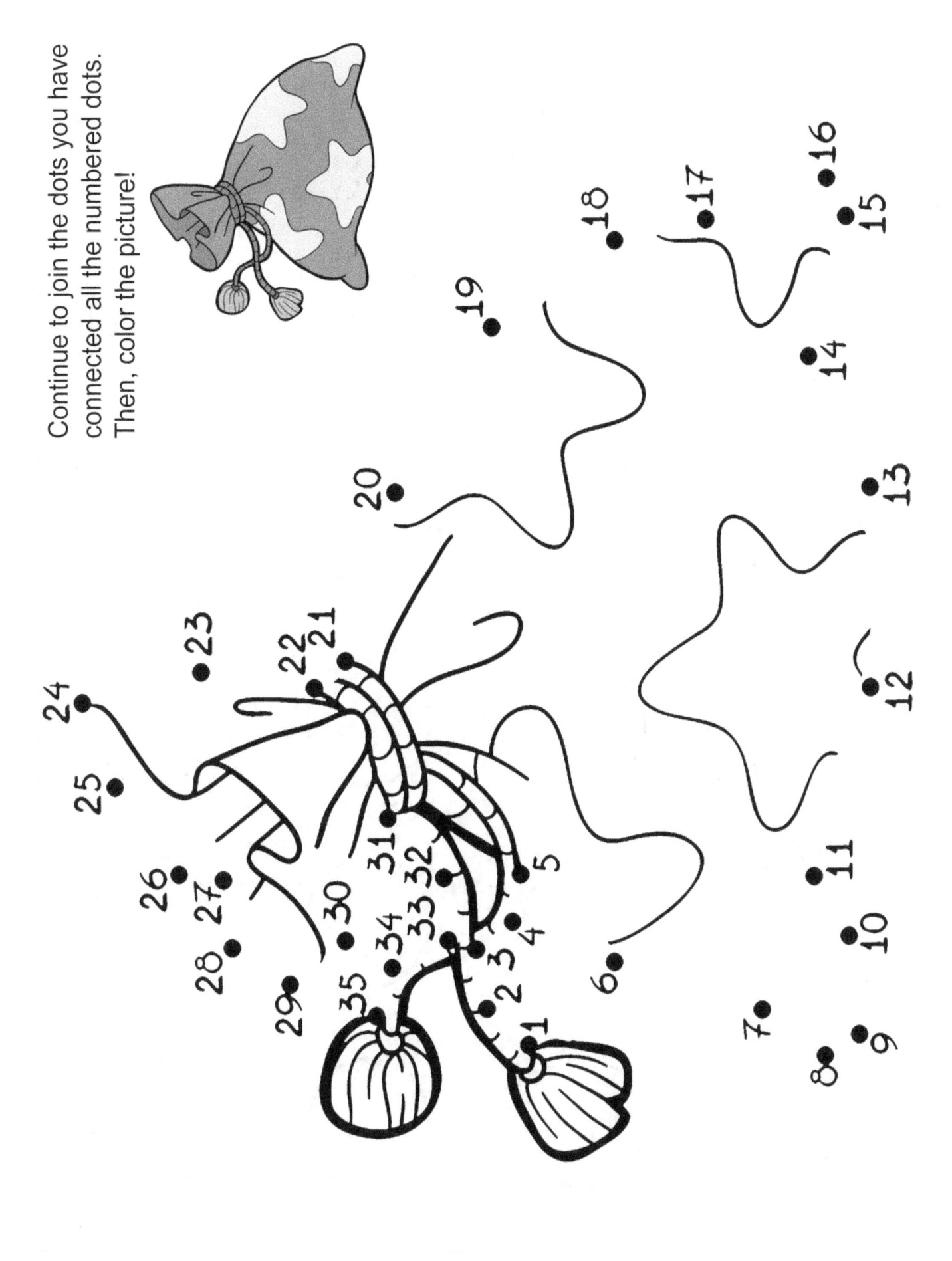

Test Your Color

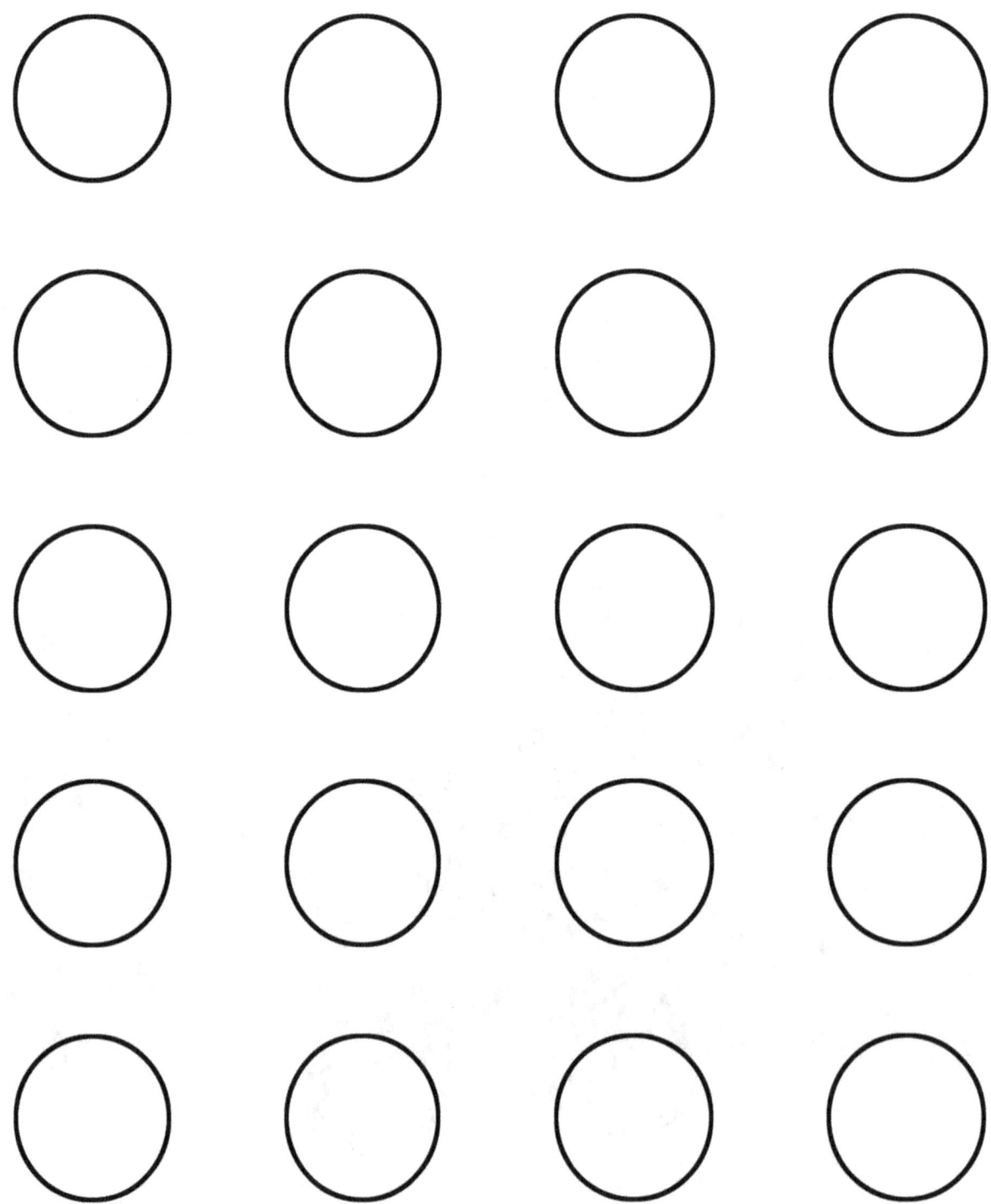

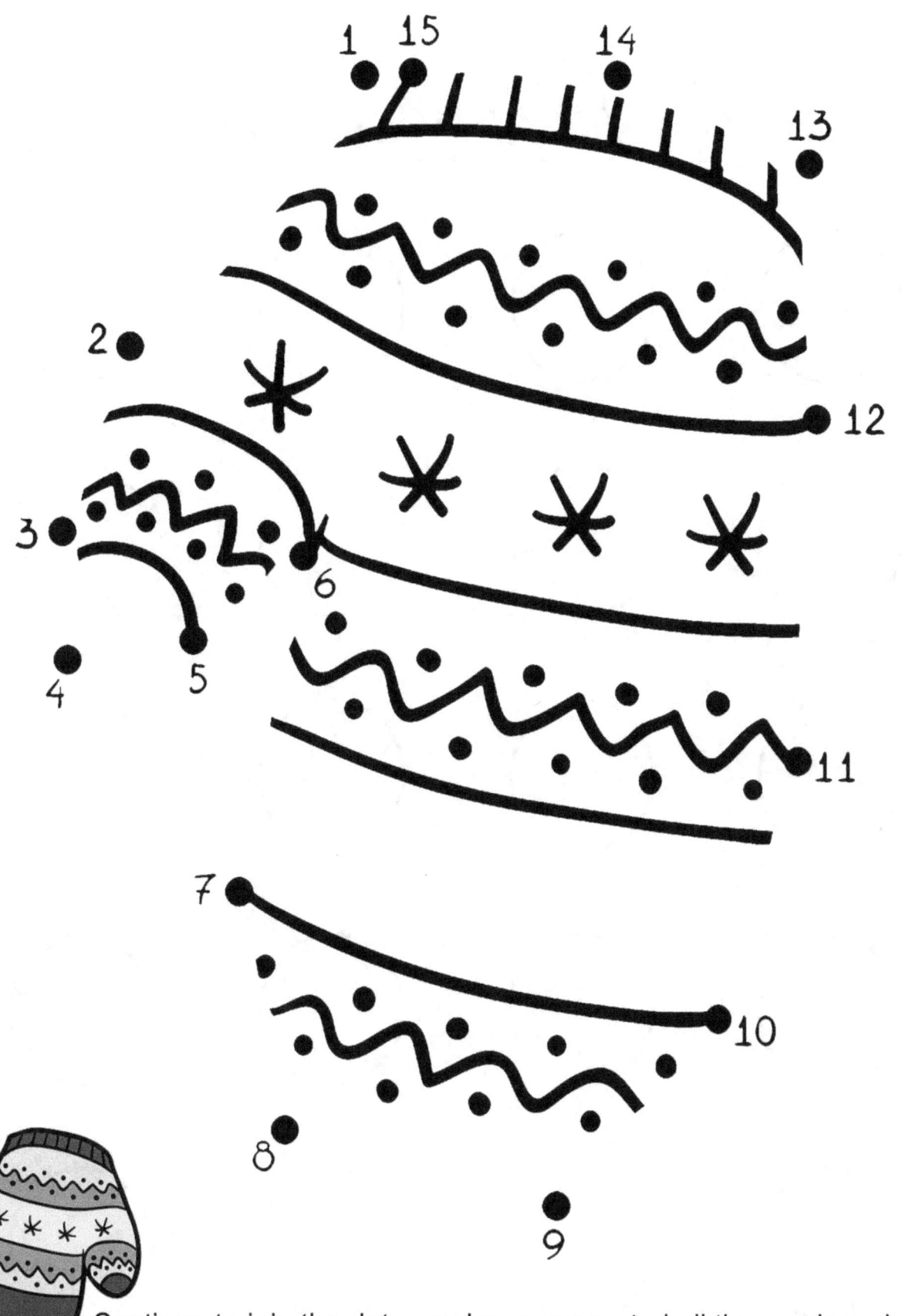

Continue to join the dots you have connected all the numbered dots. Then, color the picture!

Test Your Color

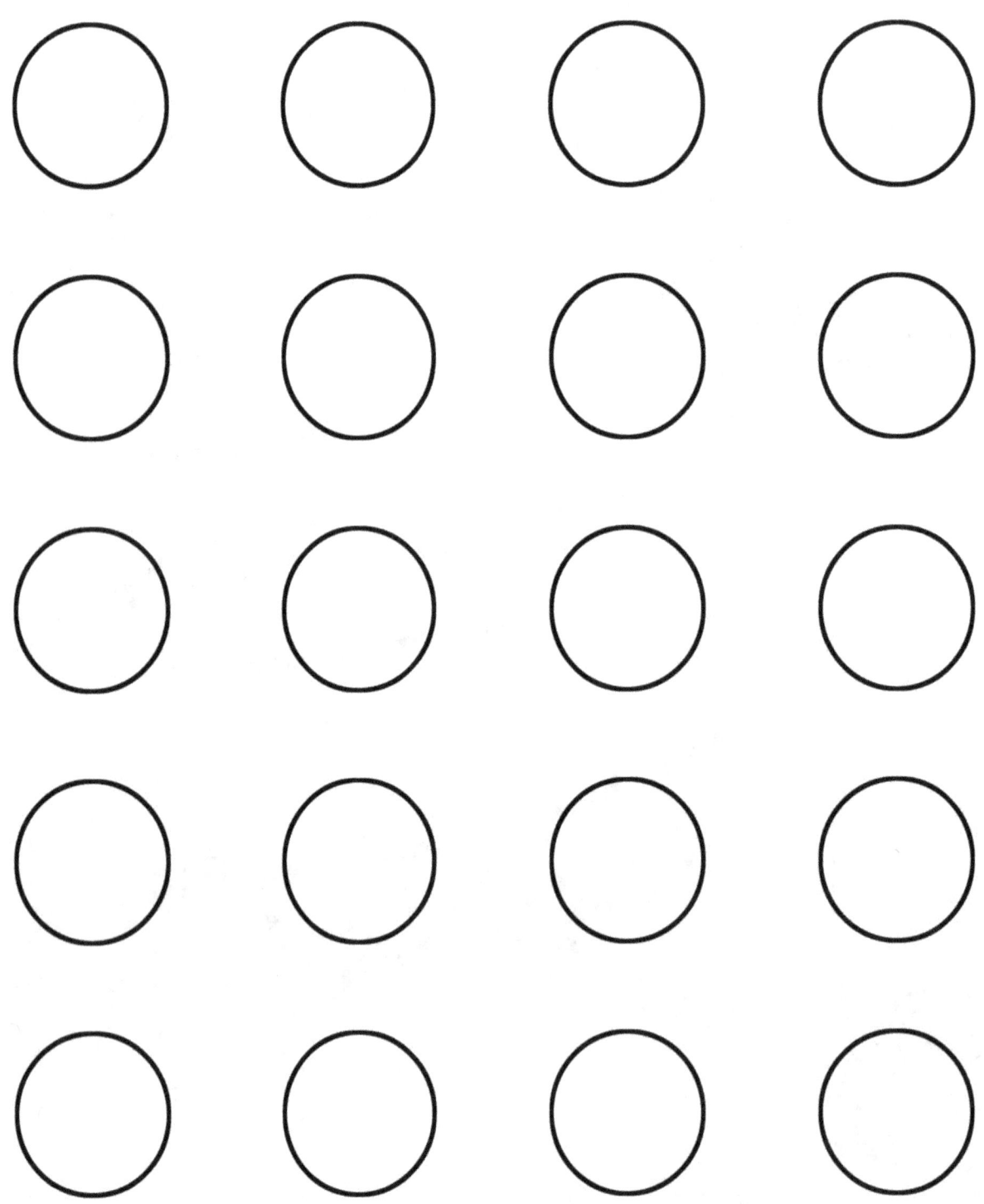

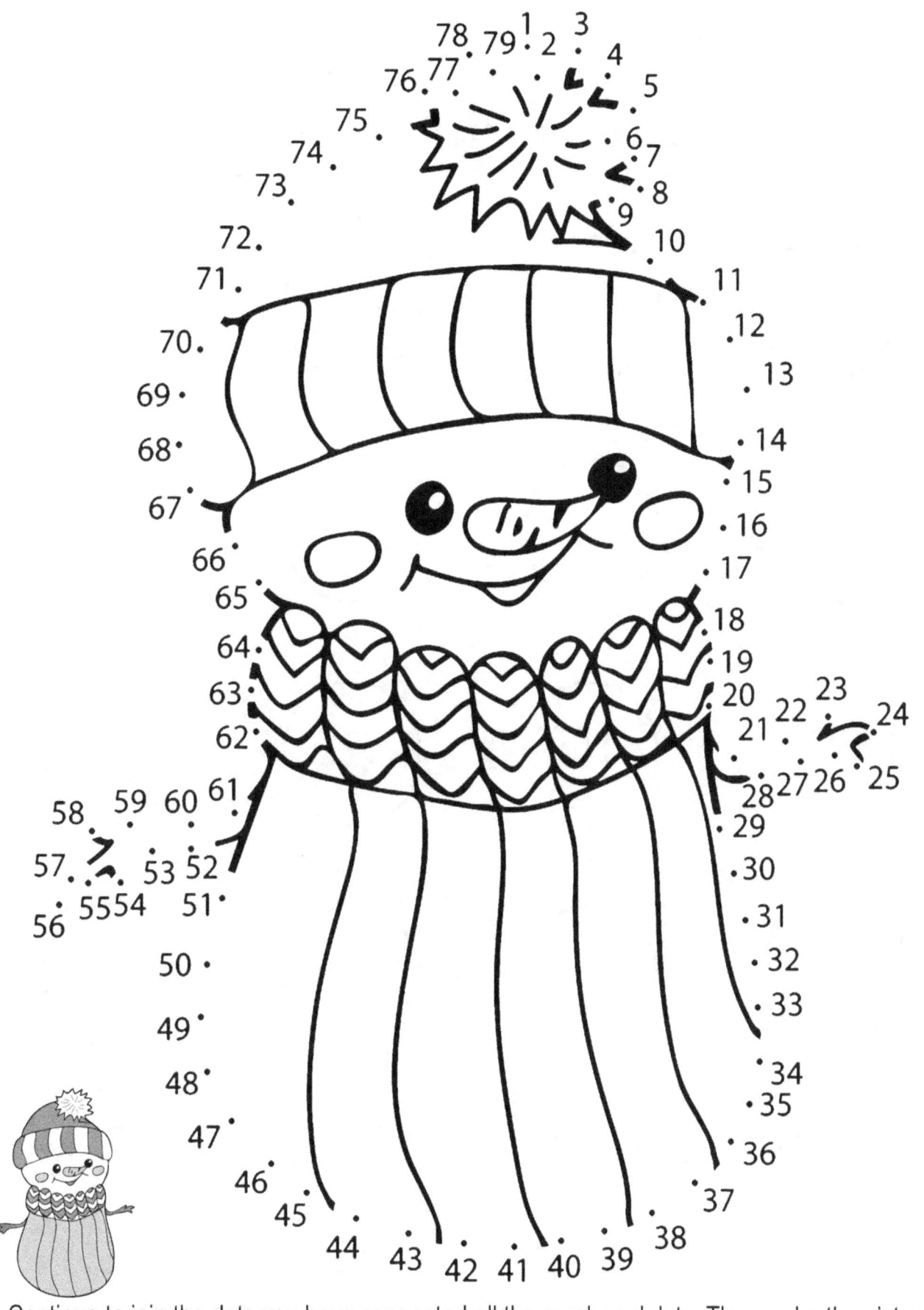

Continue to join the dots you have connected all the numbered dots. Then, color the picture!

Test Your Color

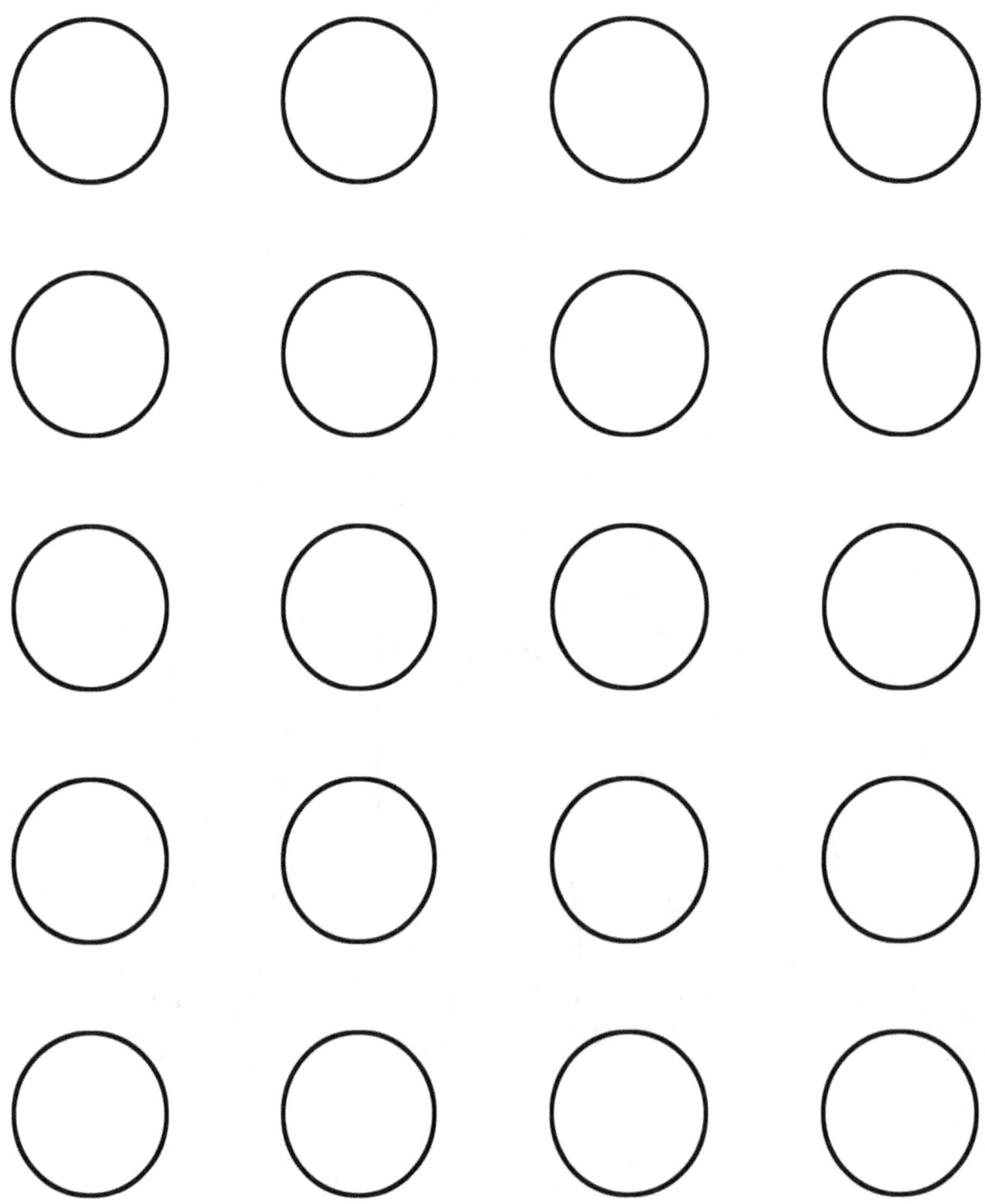

Continue to join the dots you have connected all the numbered dots. Then, color the picture!

Test Your Color

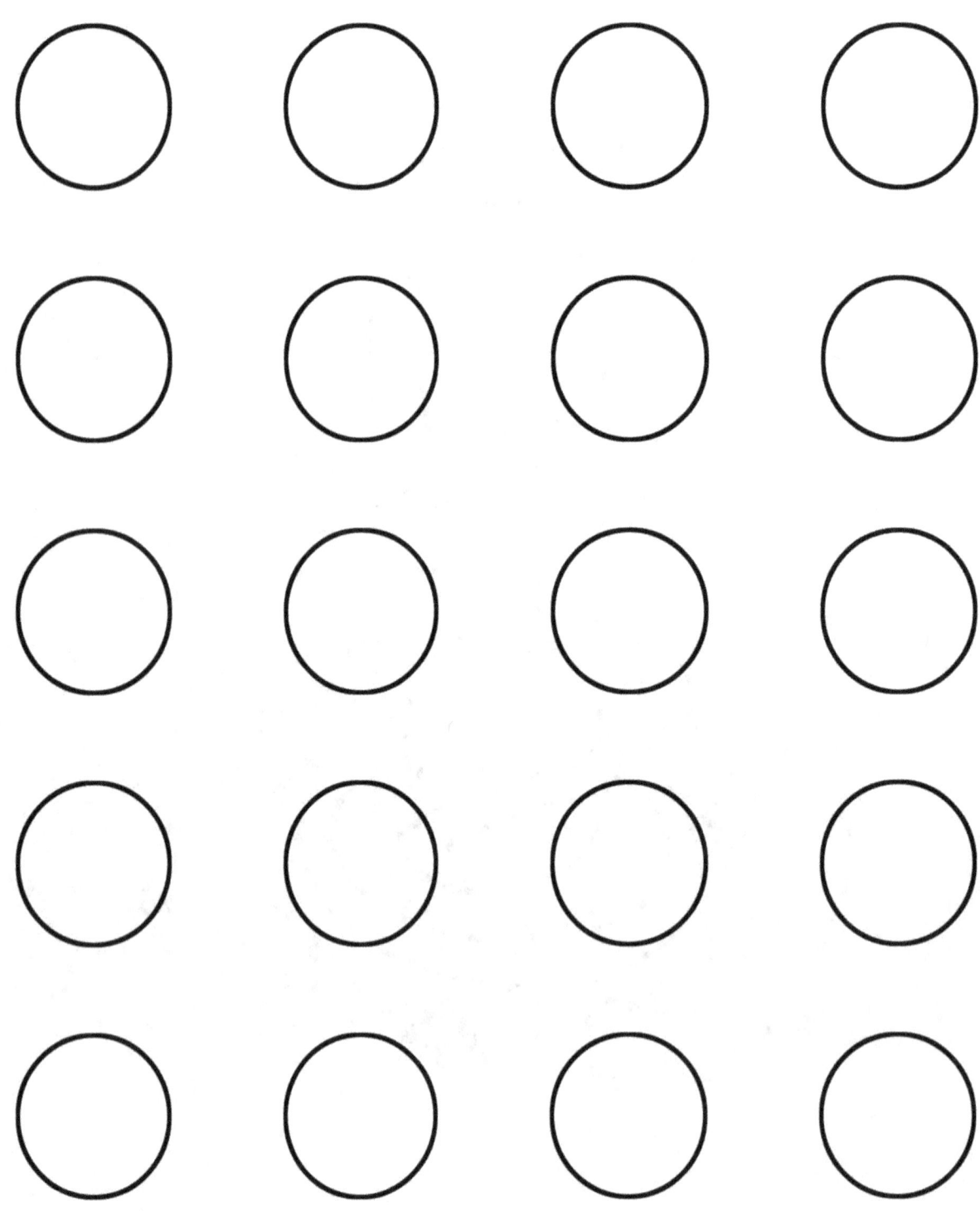

Continue to join the dots you have connected all the numbered dots. Then, color the picture!

Test Your Color

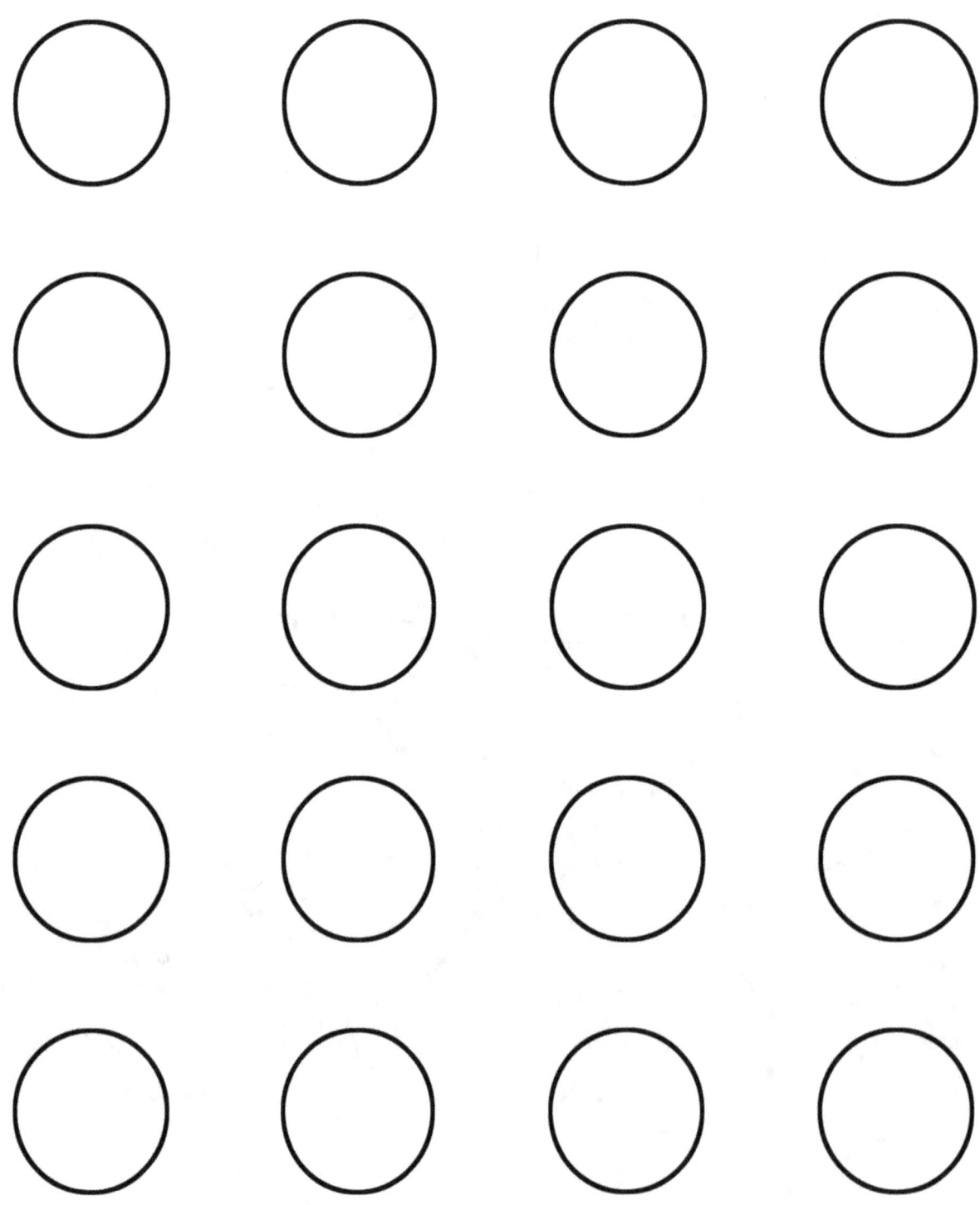

Continue to join the dots you have connected all the numbered dots. Then, color the picture!

Test Your Color

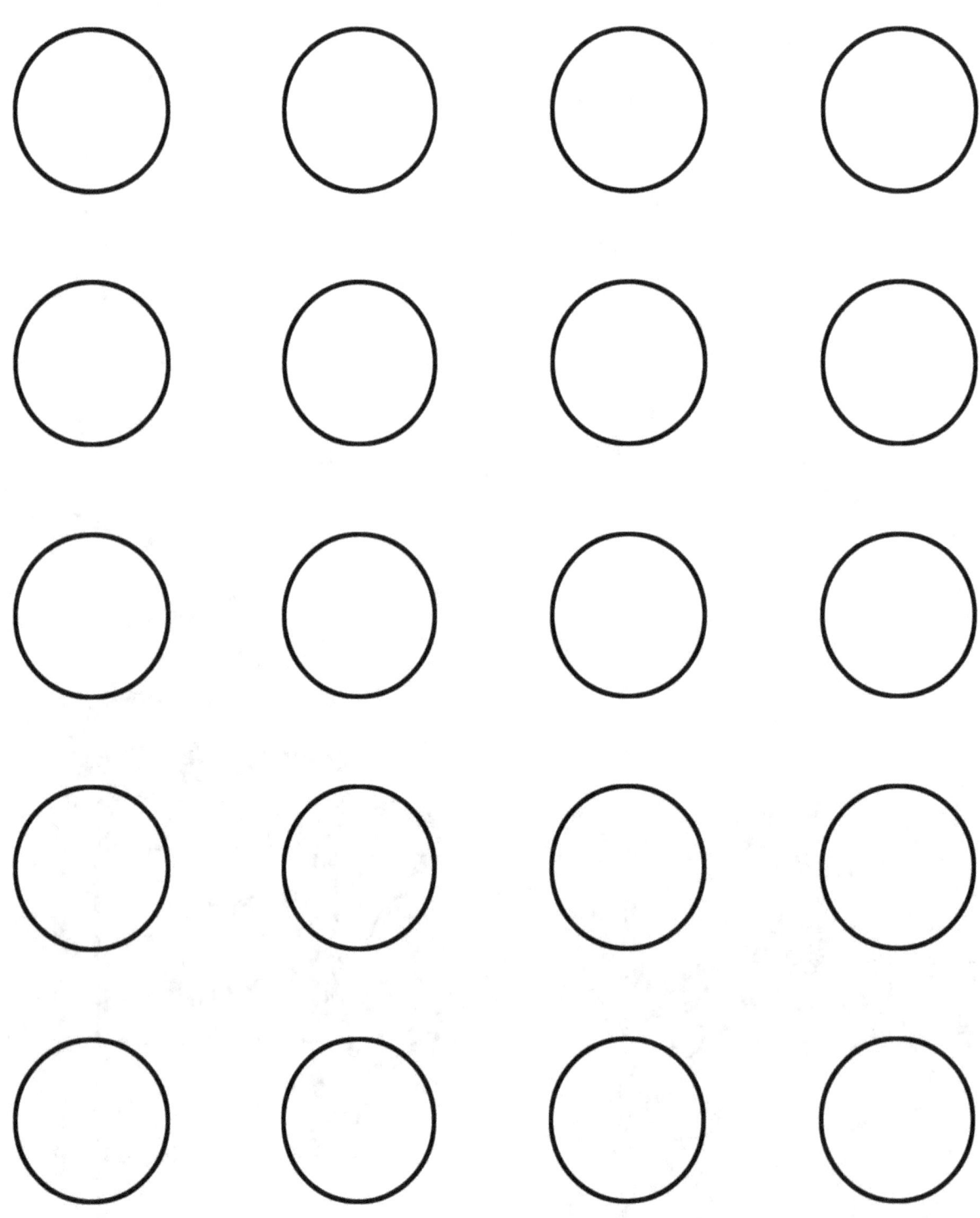

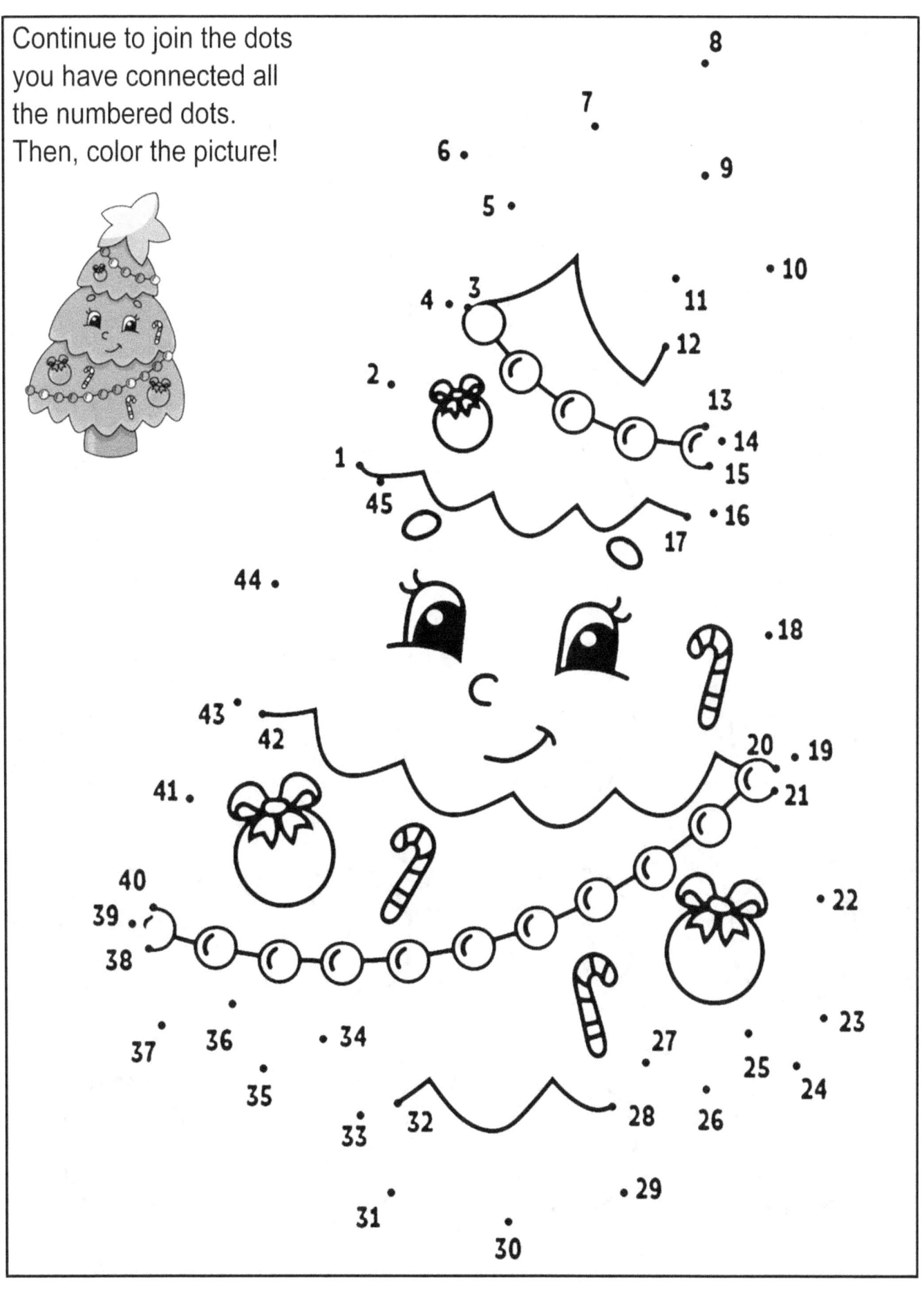

Test Your Color

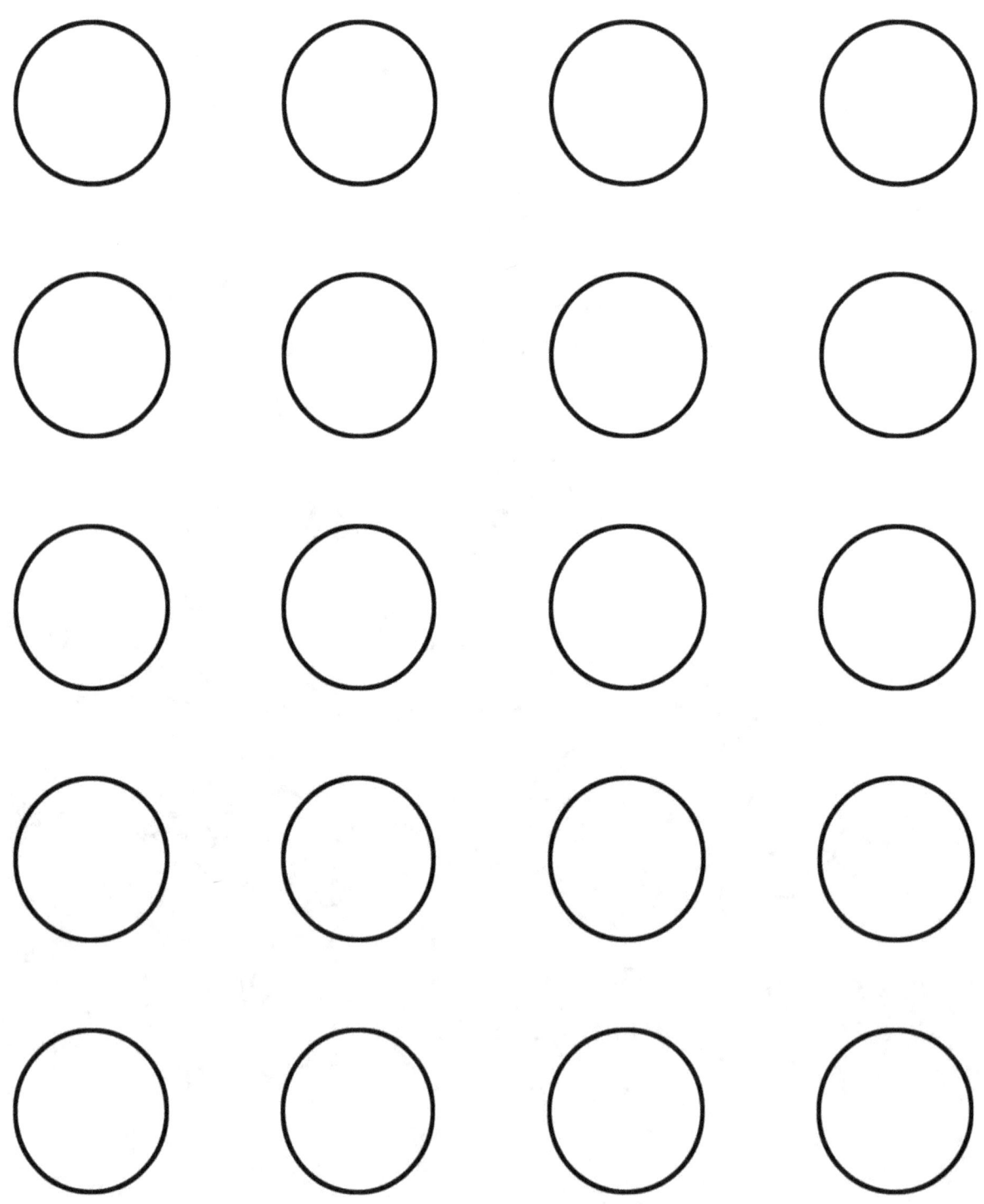

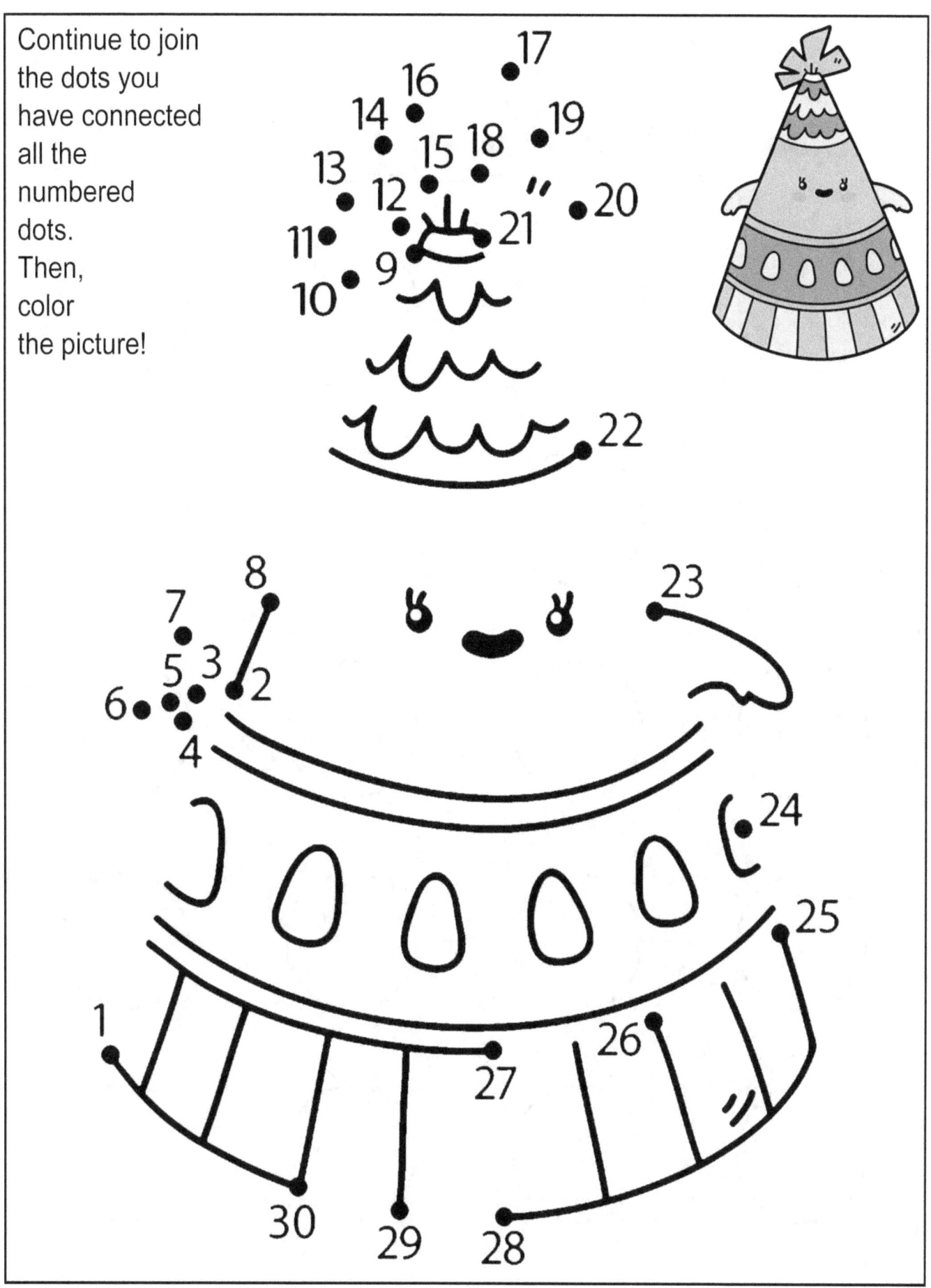

Test Your Color

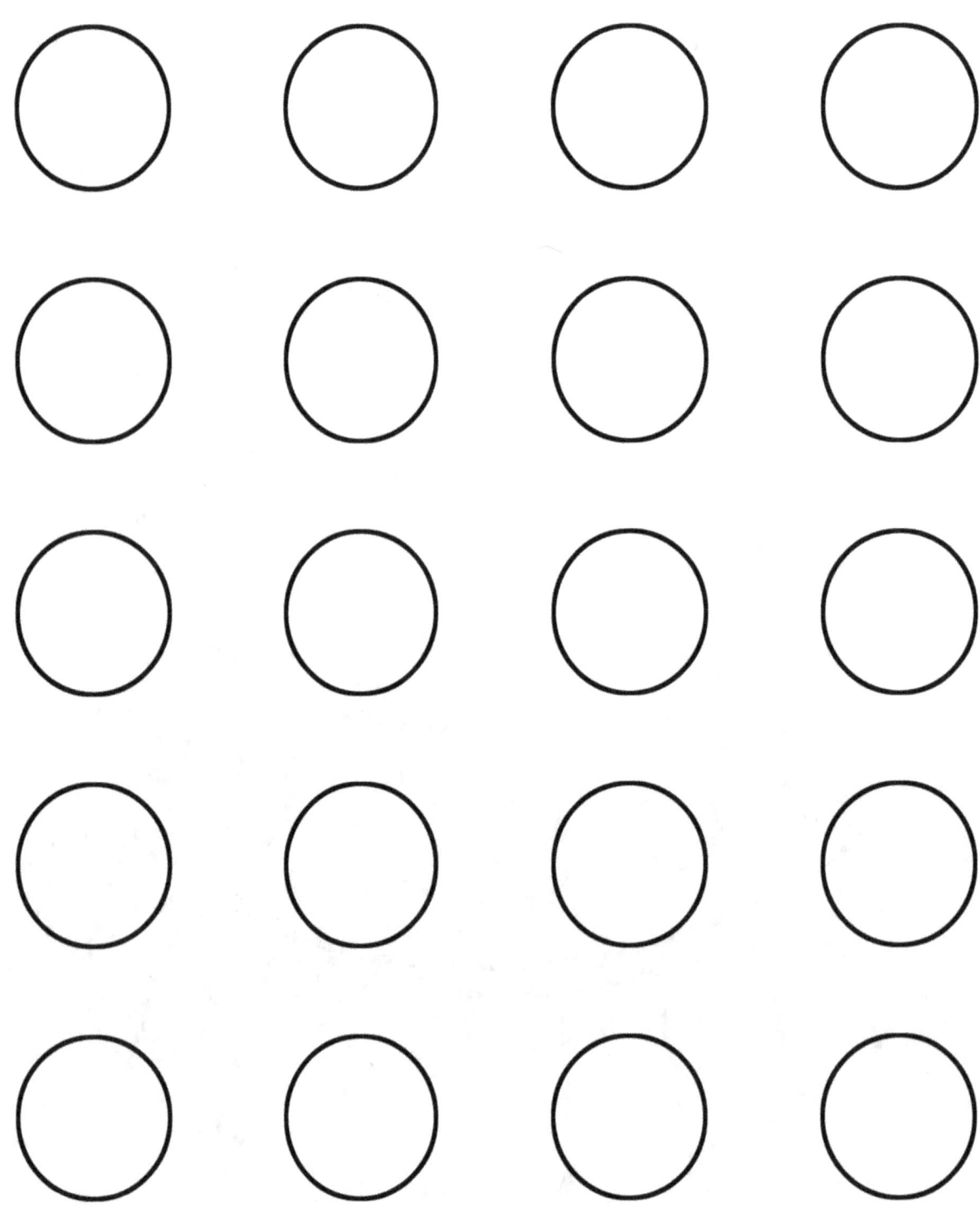

Continue to join the dots you have connected all the numbered dots. Then, color the picture!

Test Your Color

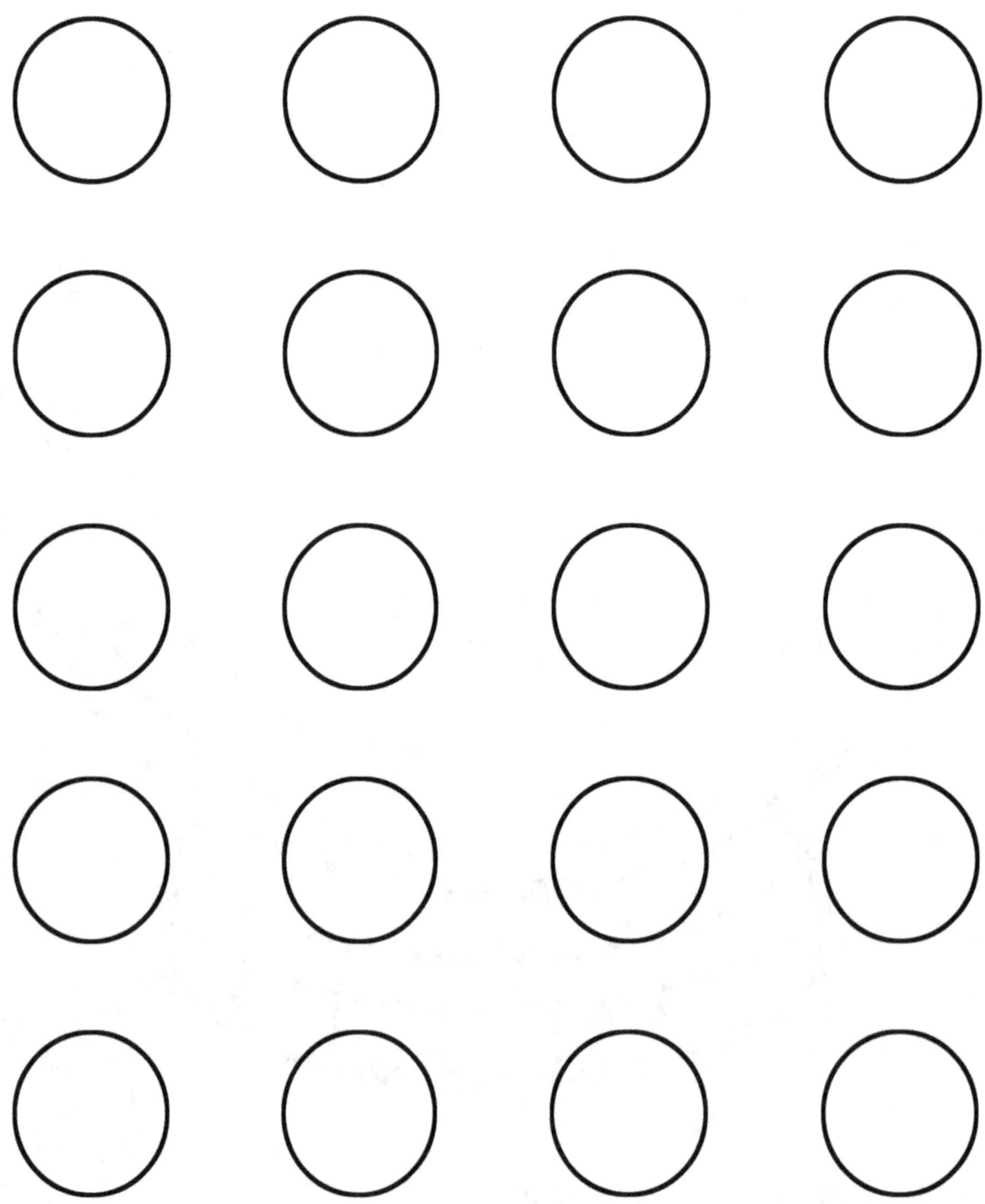

Continue to join the dots you have connected all the numbered dots. Then, color the picture!

Test Your Color

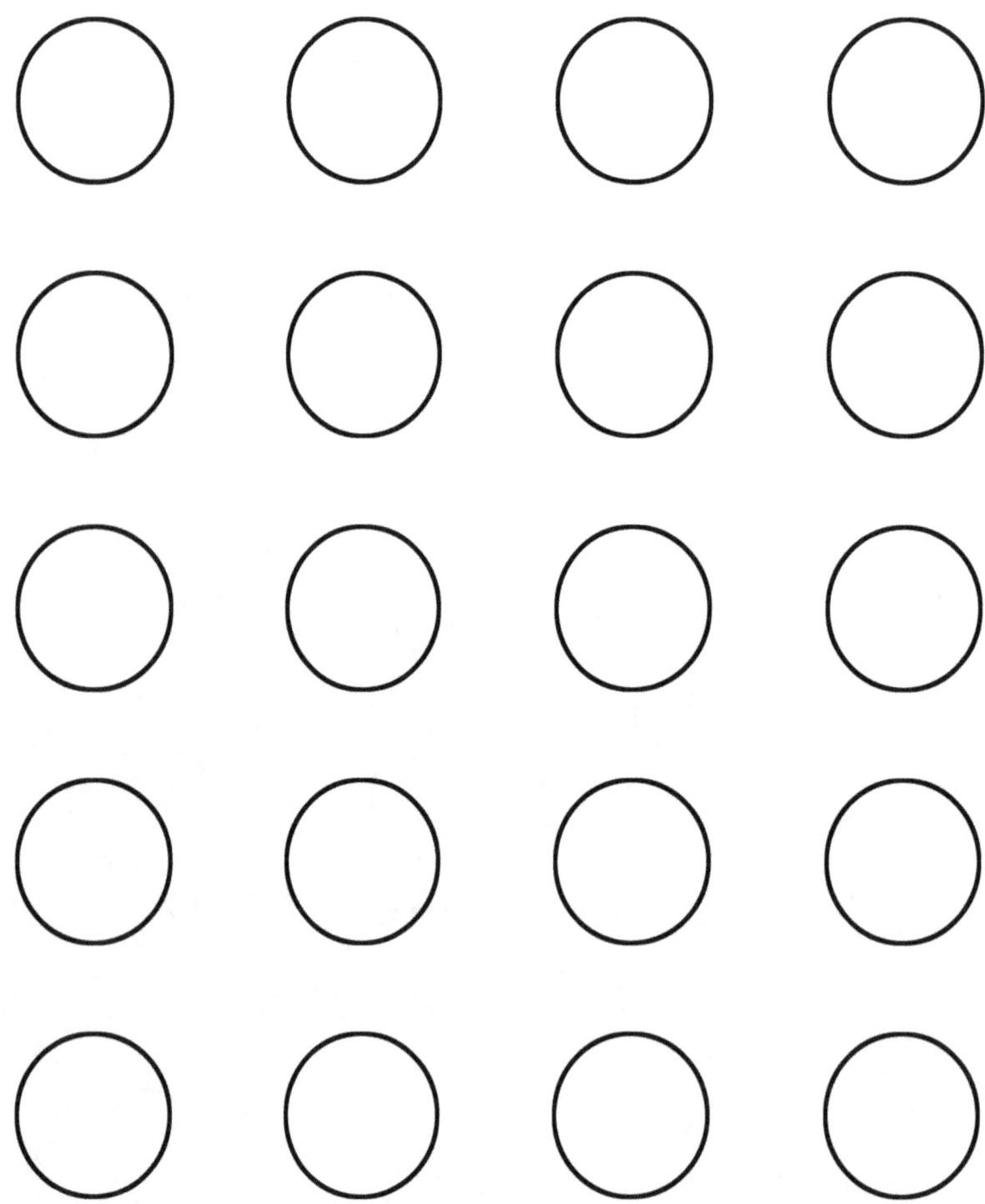

Continue to join the dots you have connected all the numbered dots. Then, color the picture!

Test Your Color

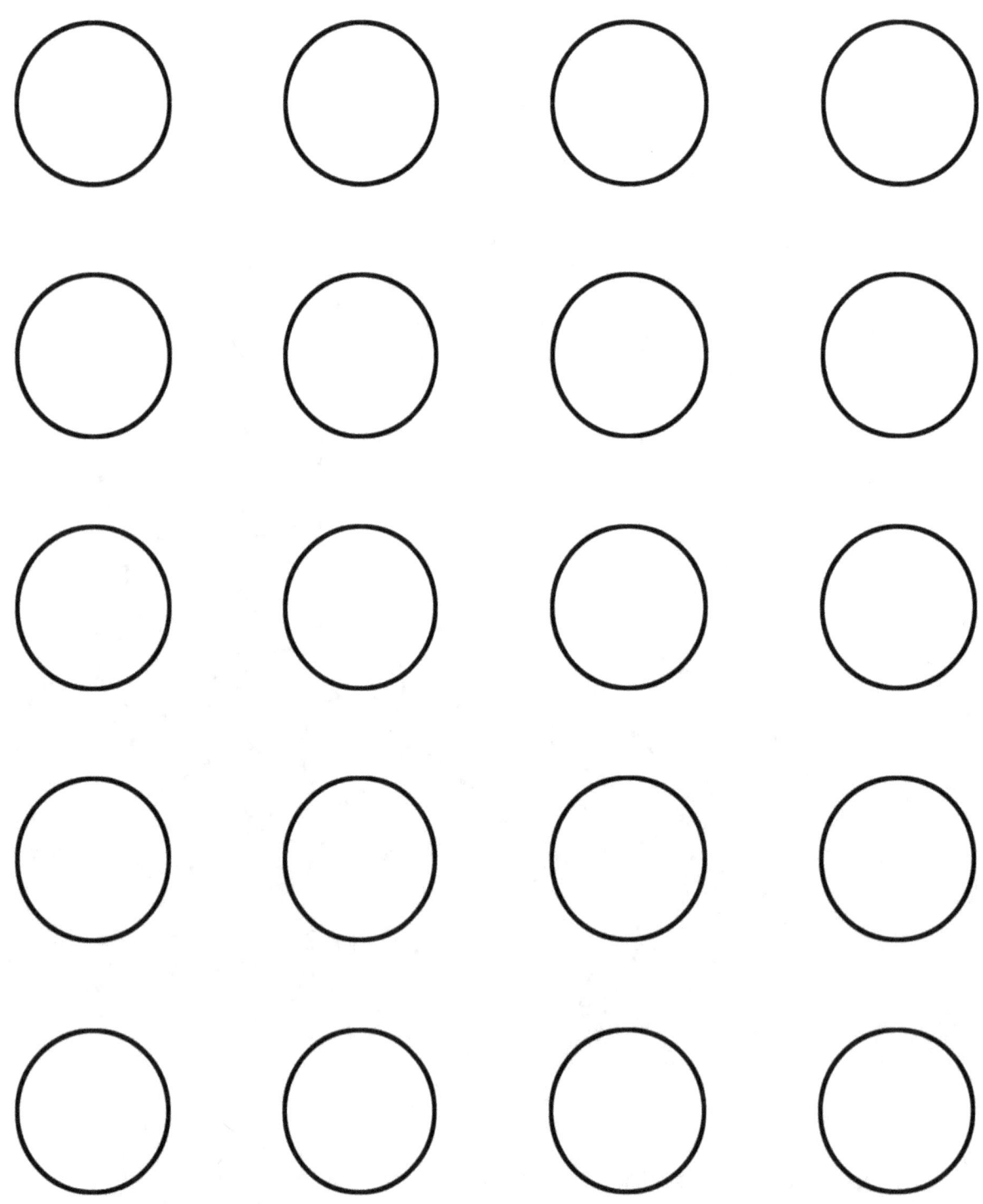

Continue to join the dots you have connected all the numbered dots. Then, color the picture!

Test Your Color

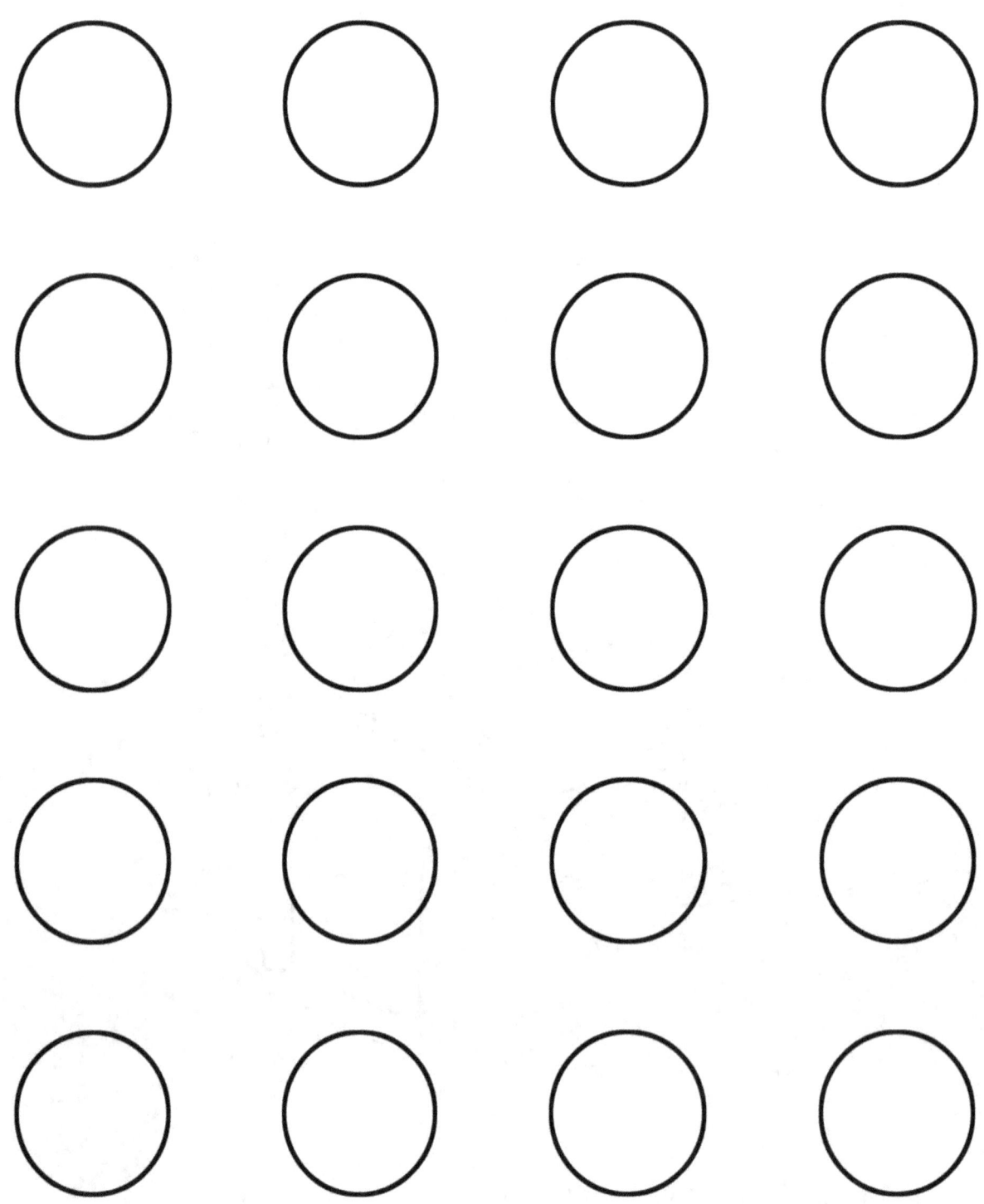

Continue to join the dots you have connected all the numbered dots. Then, color the picture!

Test Your Color

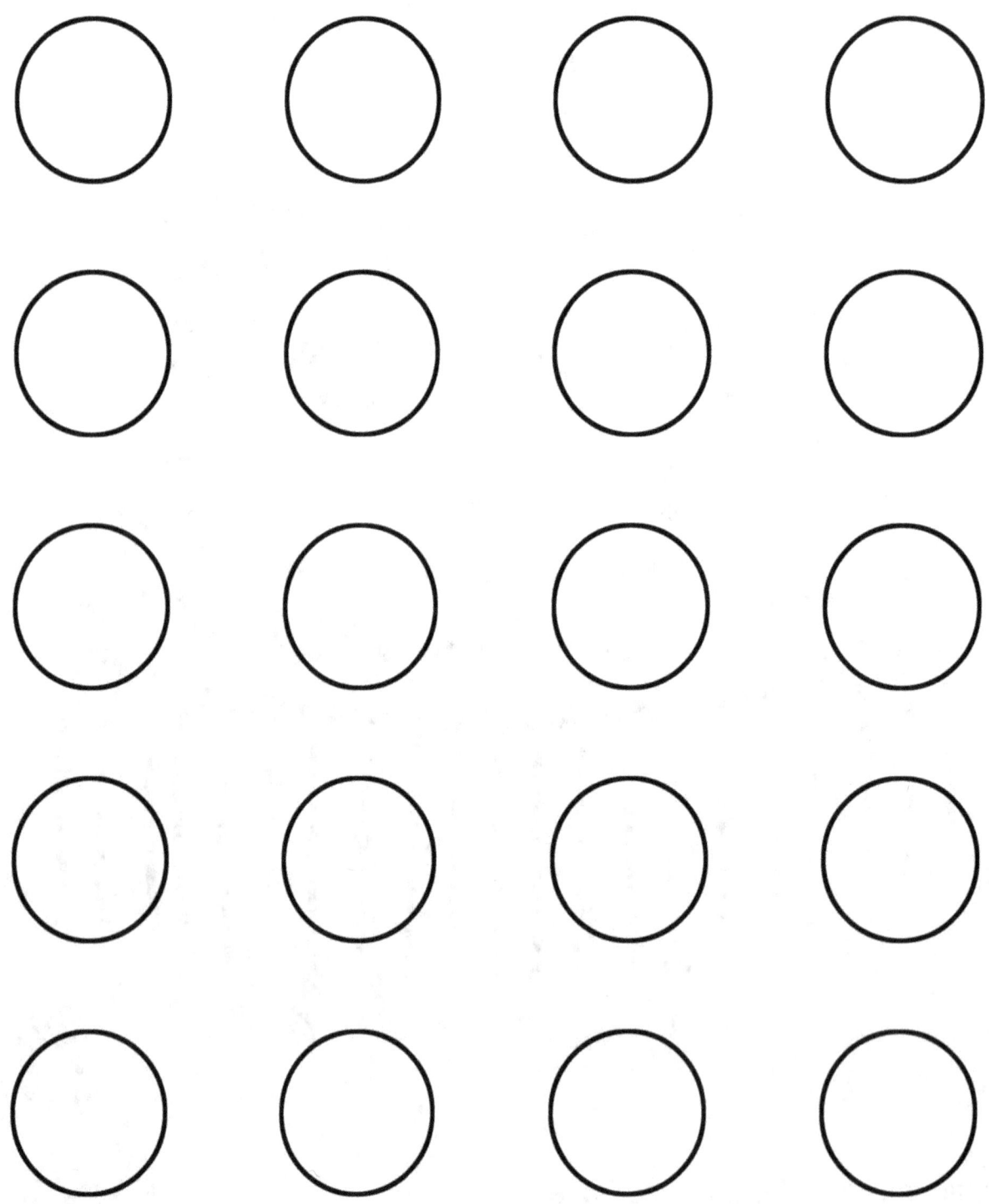

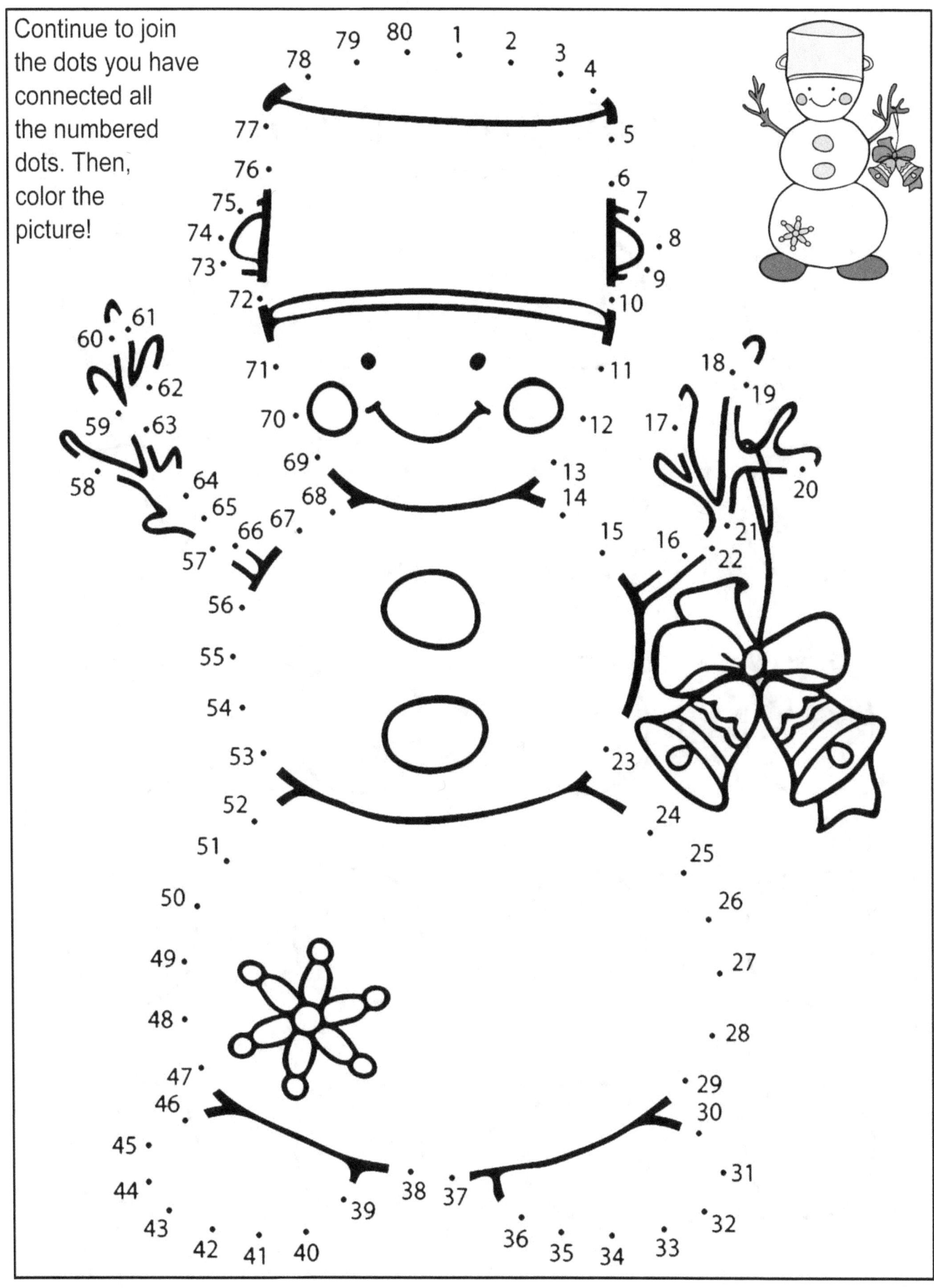

Test Your Color

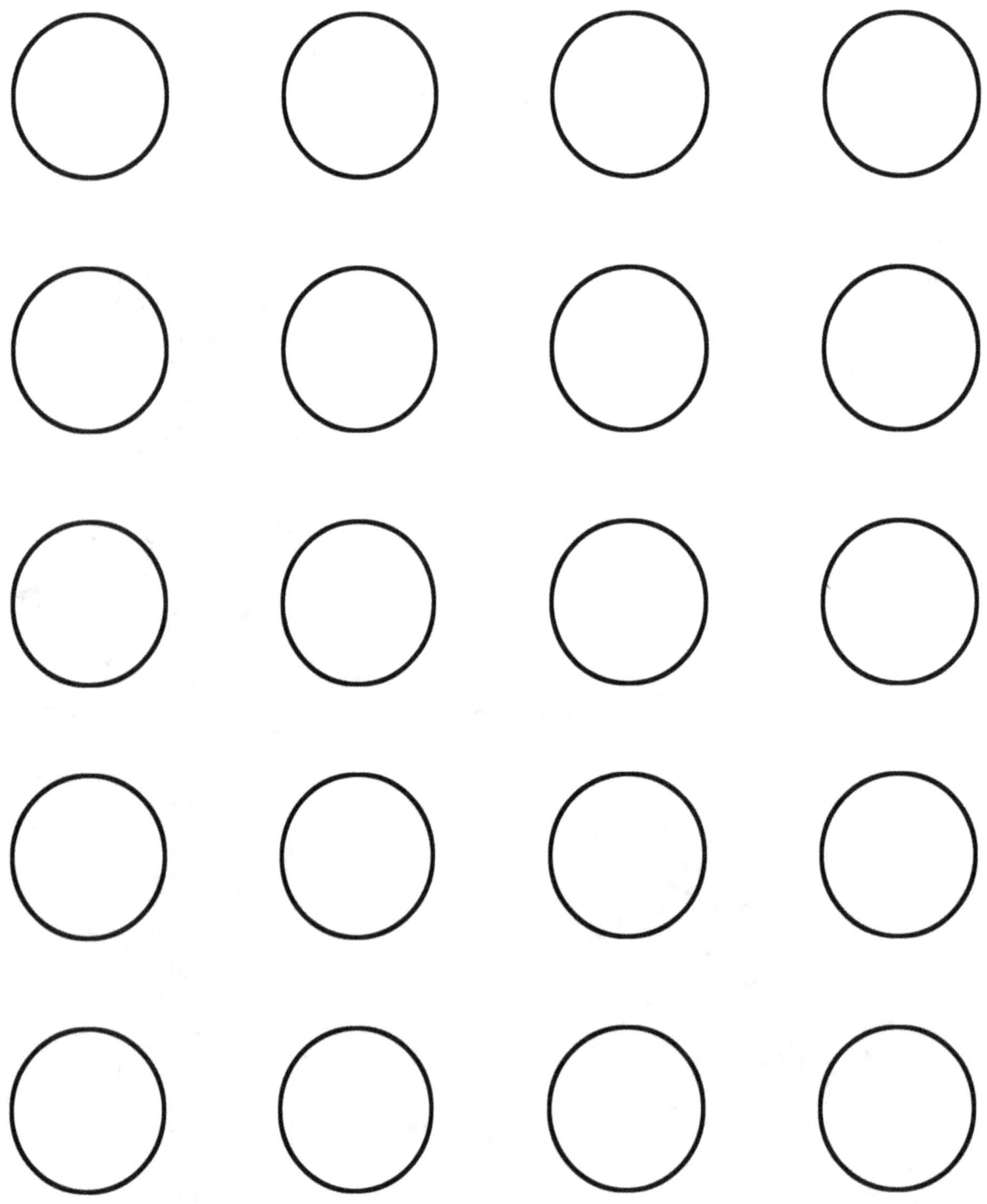

Test Your Color

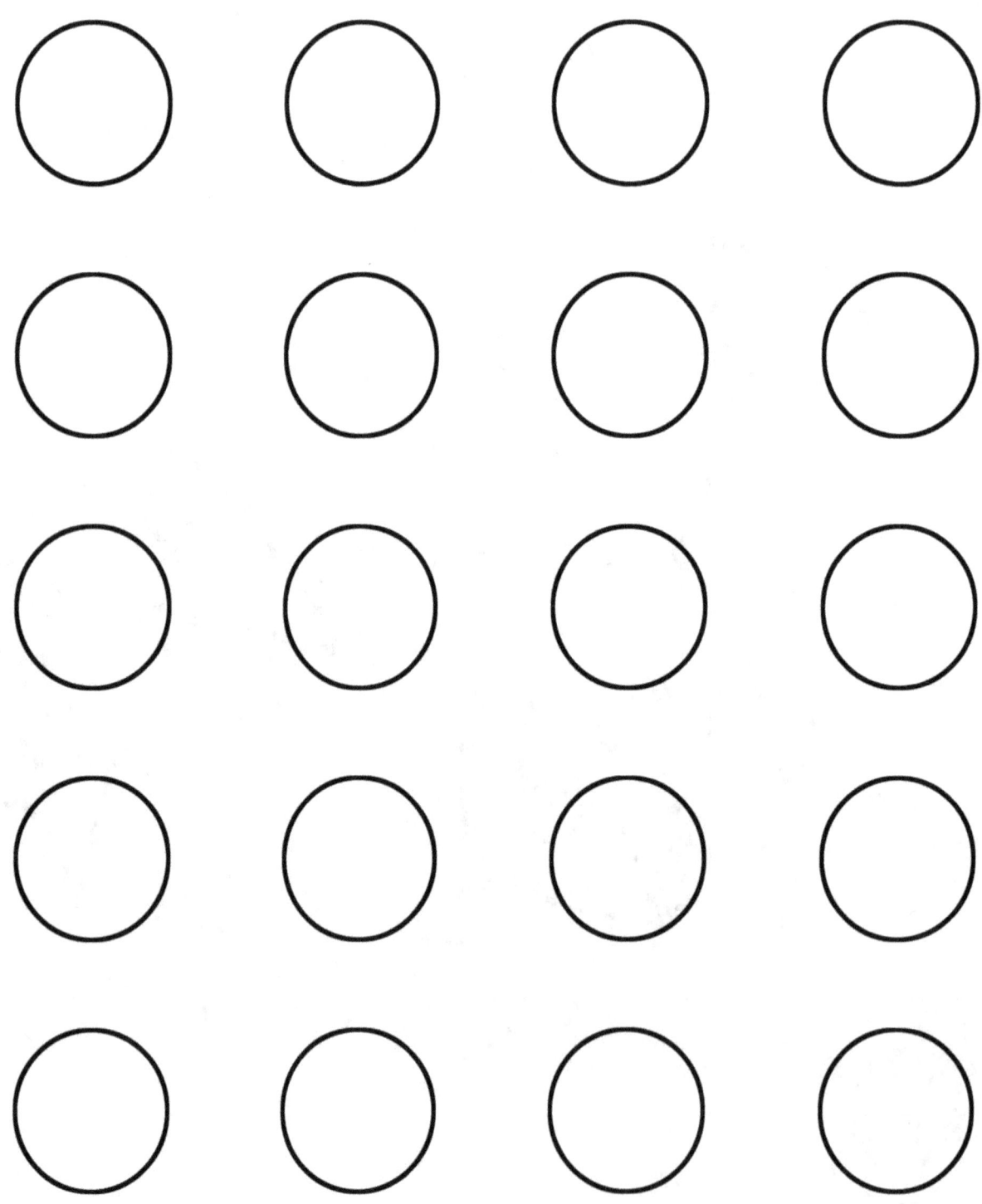

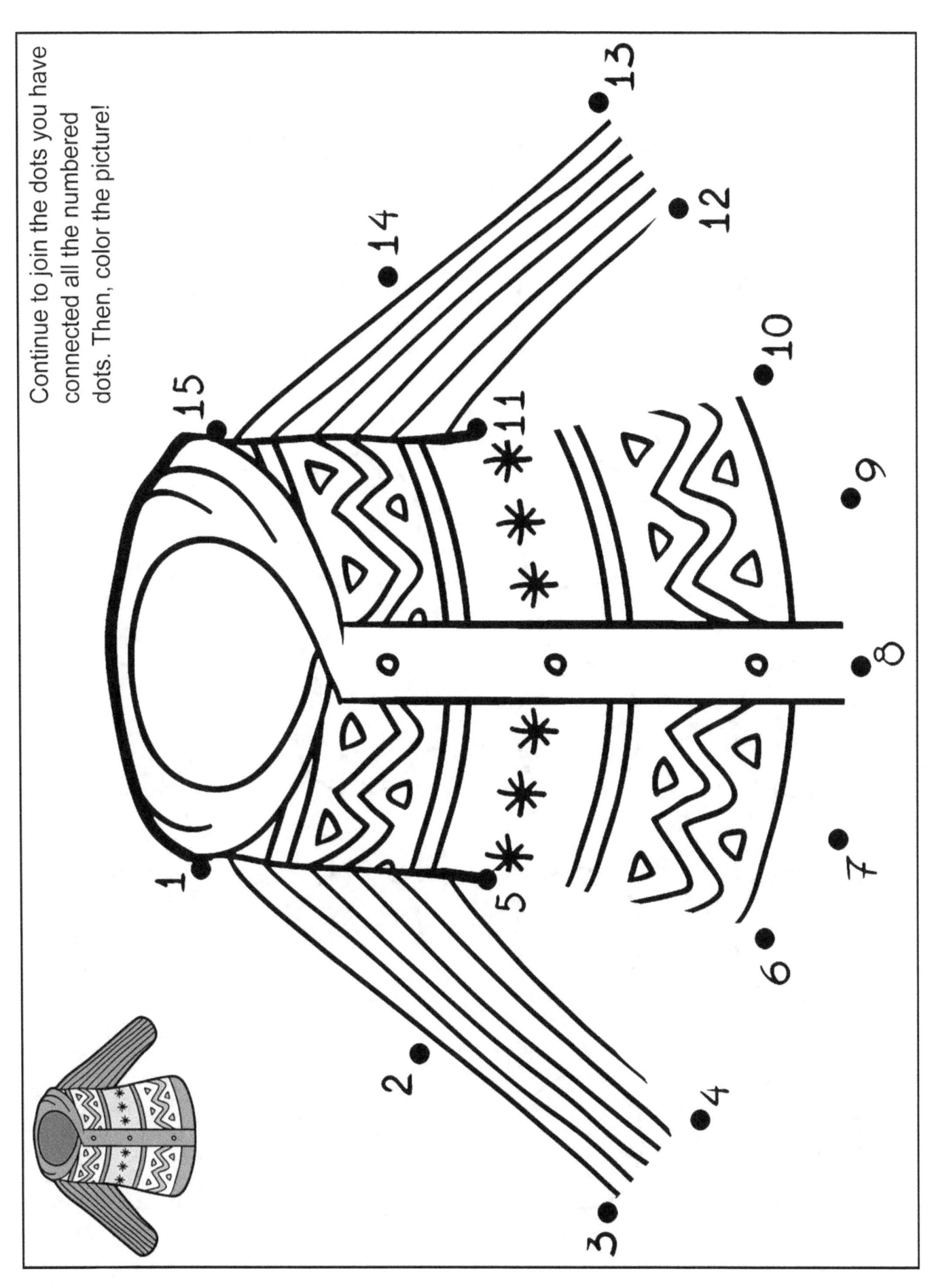

Test Your Color

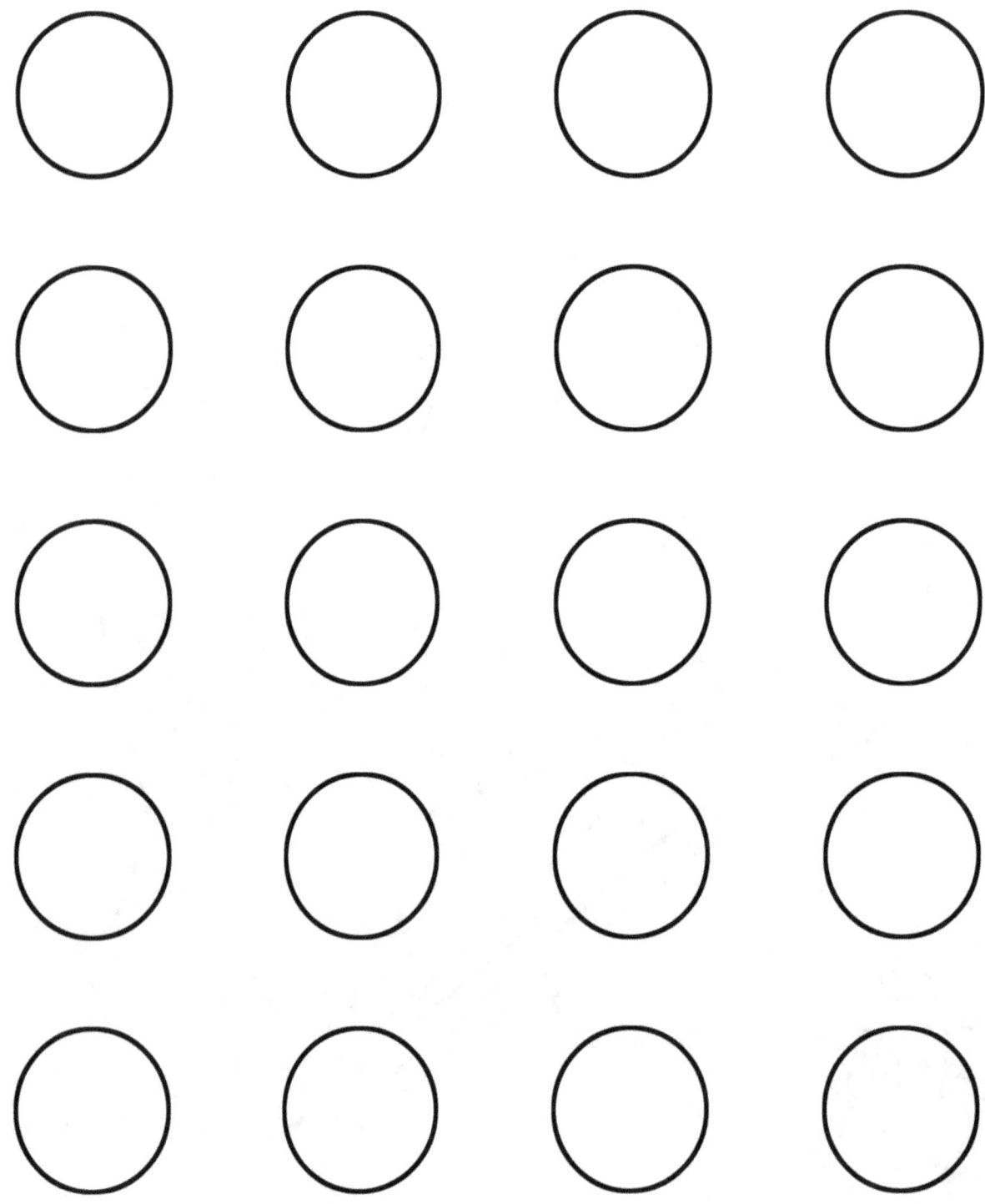

Continue to join the dots you have connected all the numbered dots. Then, color the picture!

Test Your Color

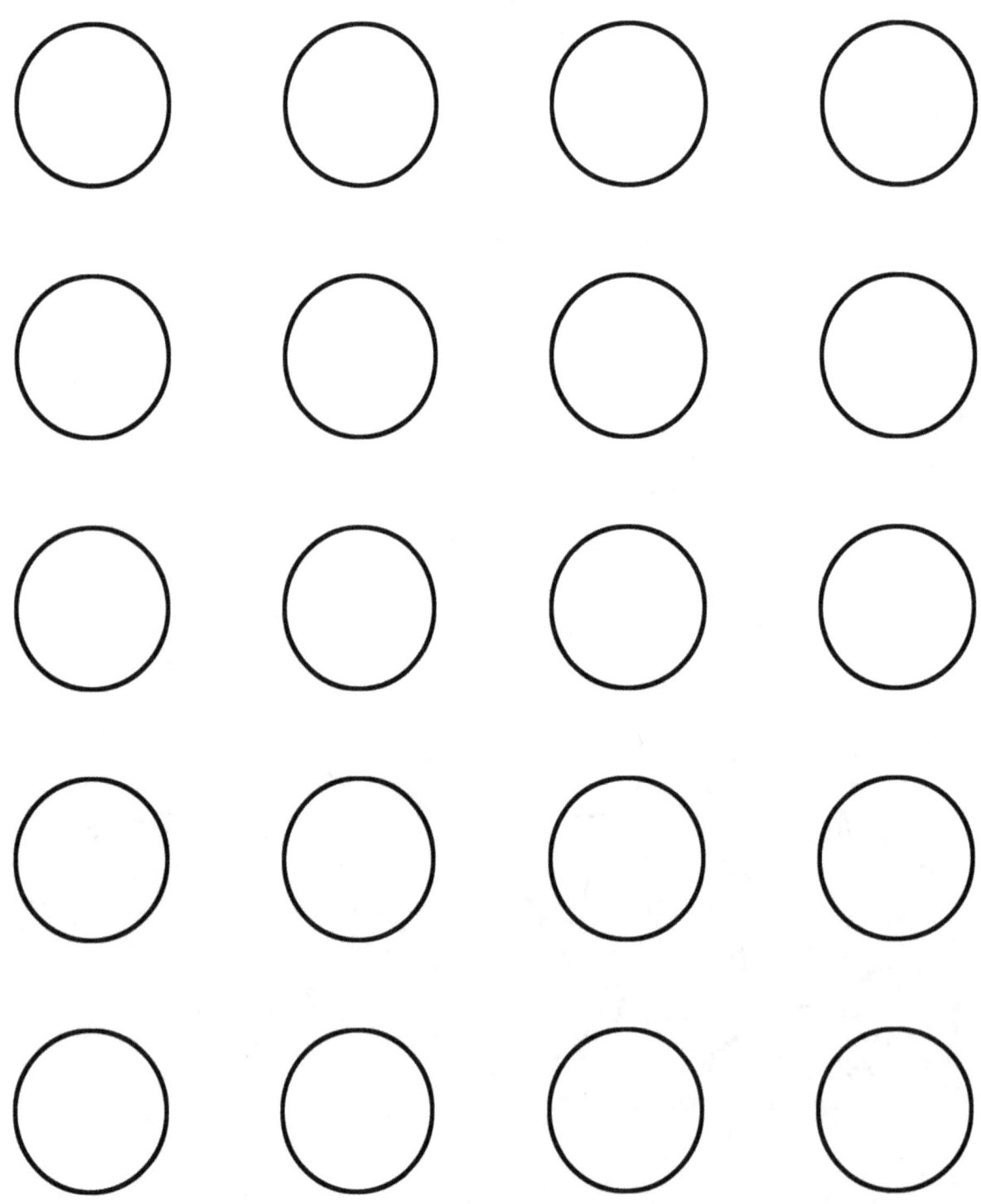

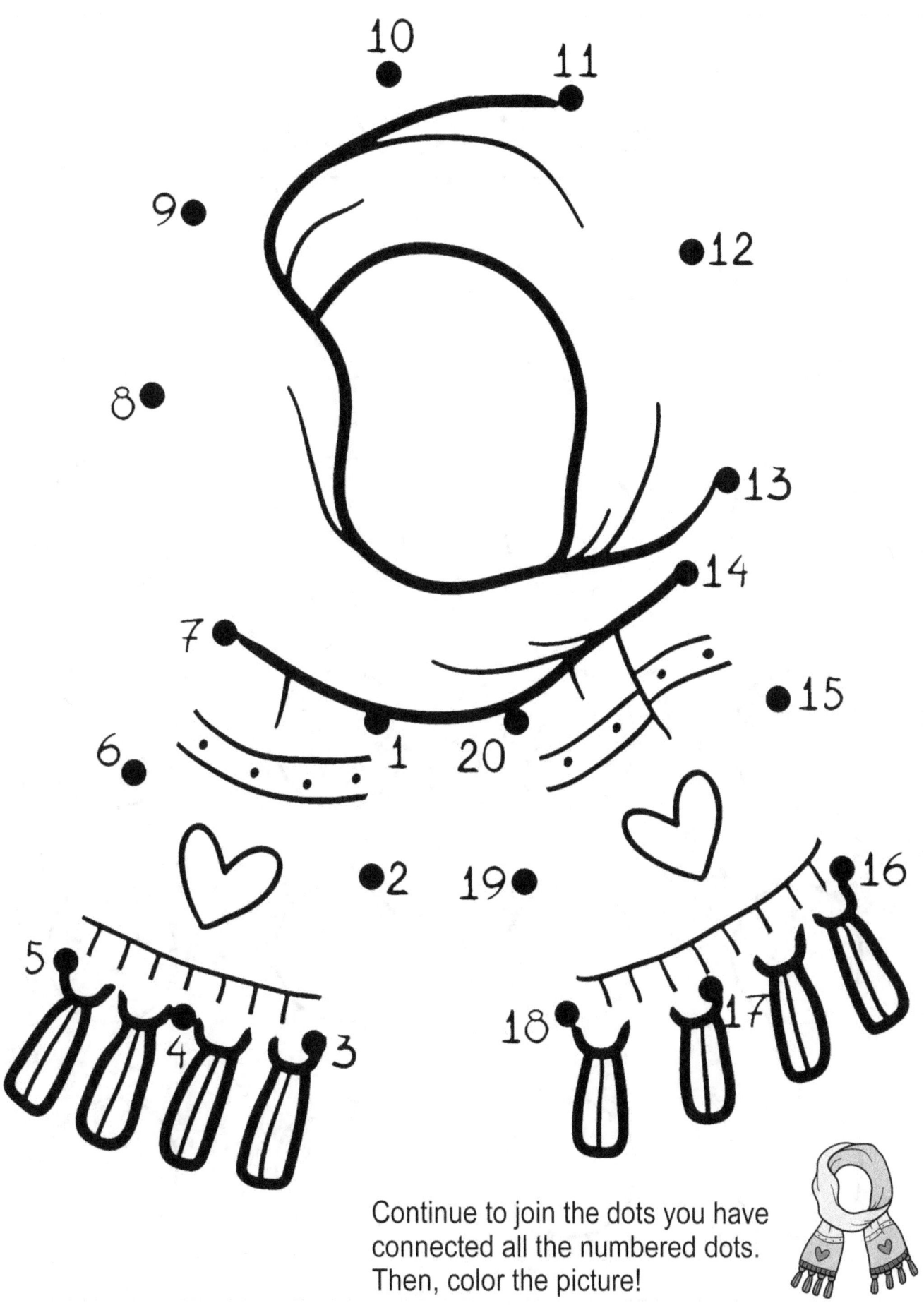

Continue to join the dots you have connected all the numbered dots. Then, color the picture!

Test Your Color

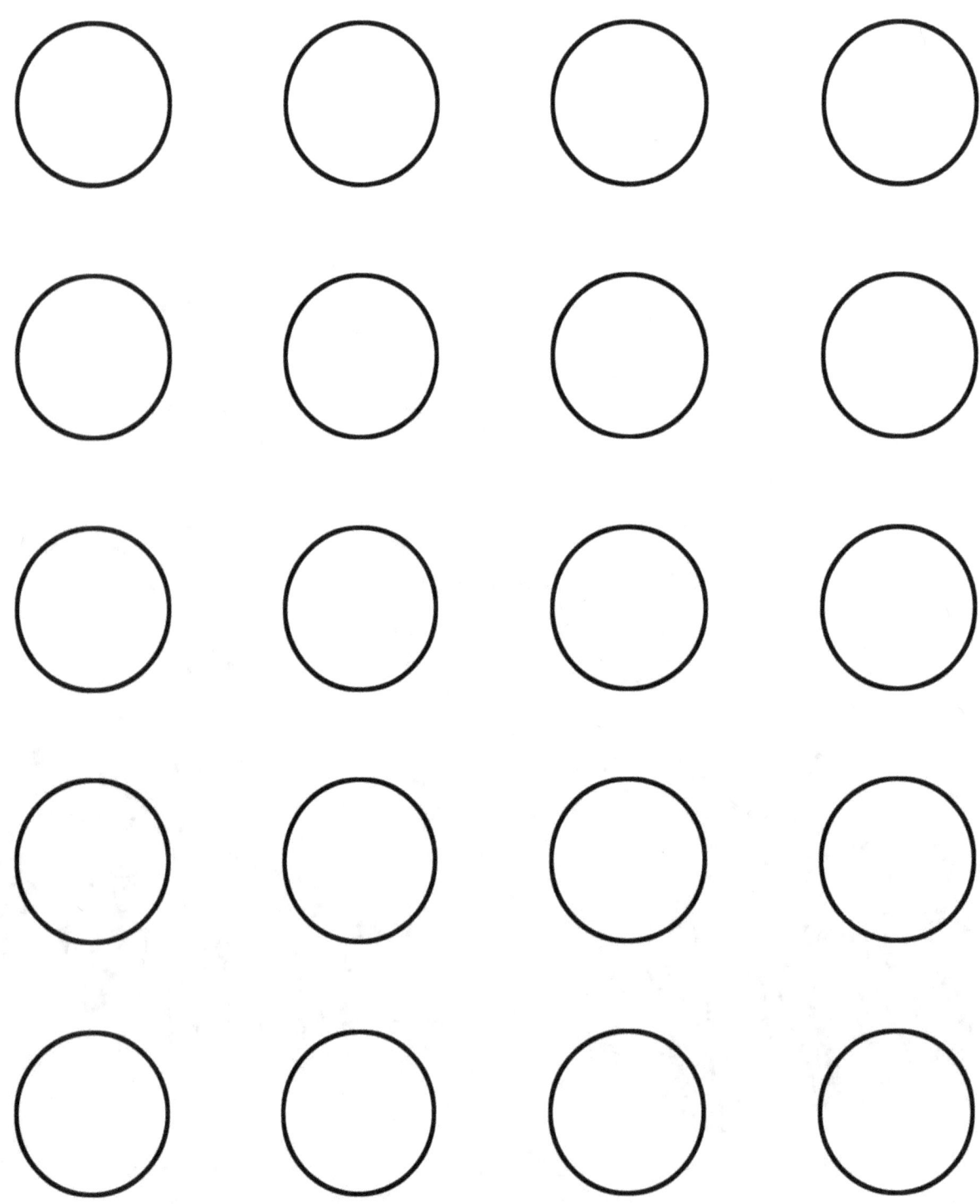

Continue to join the dots you have connected all the numbered dots.
Then, color the picture!

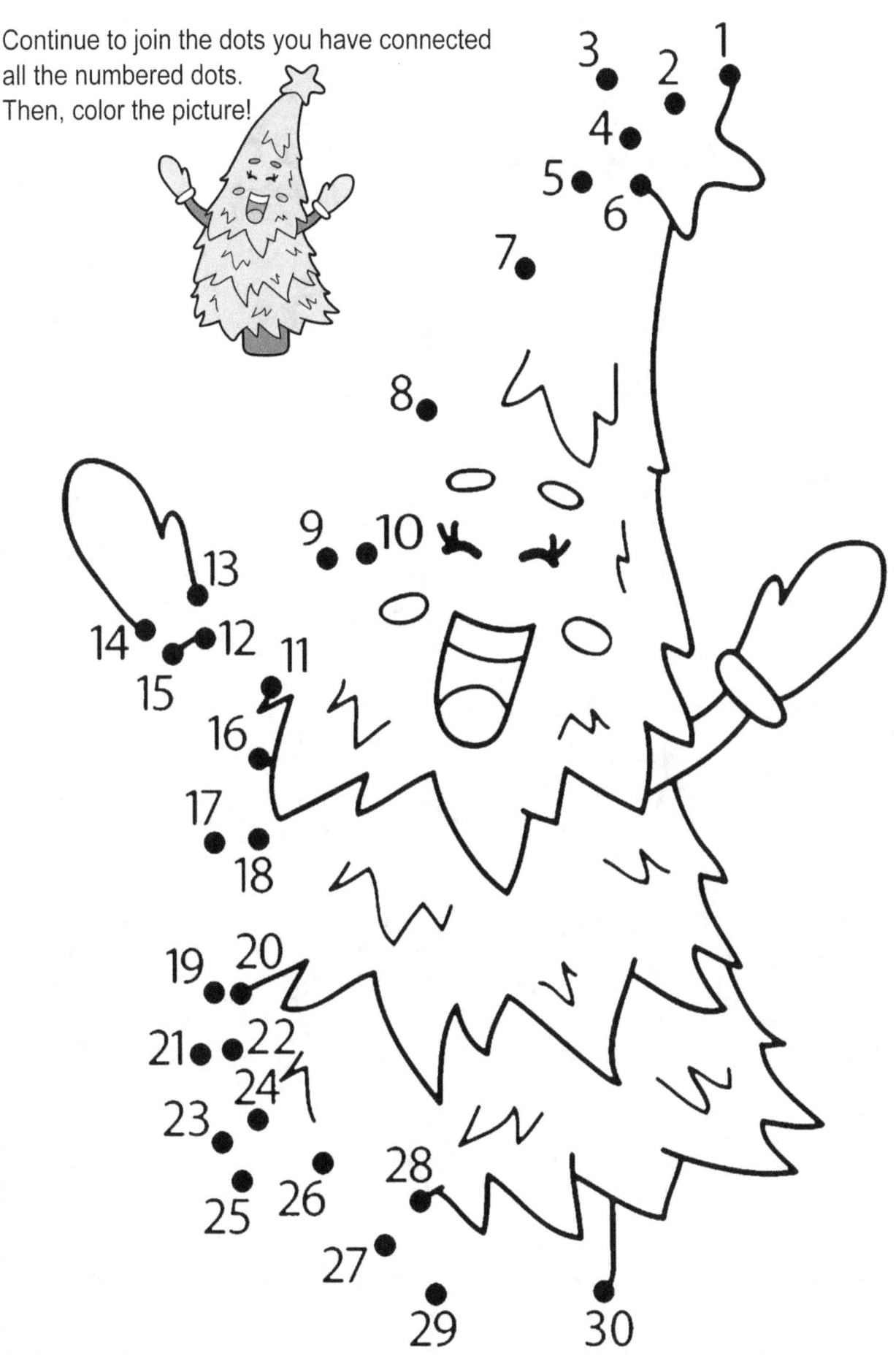

Test Your Color

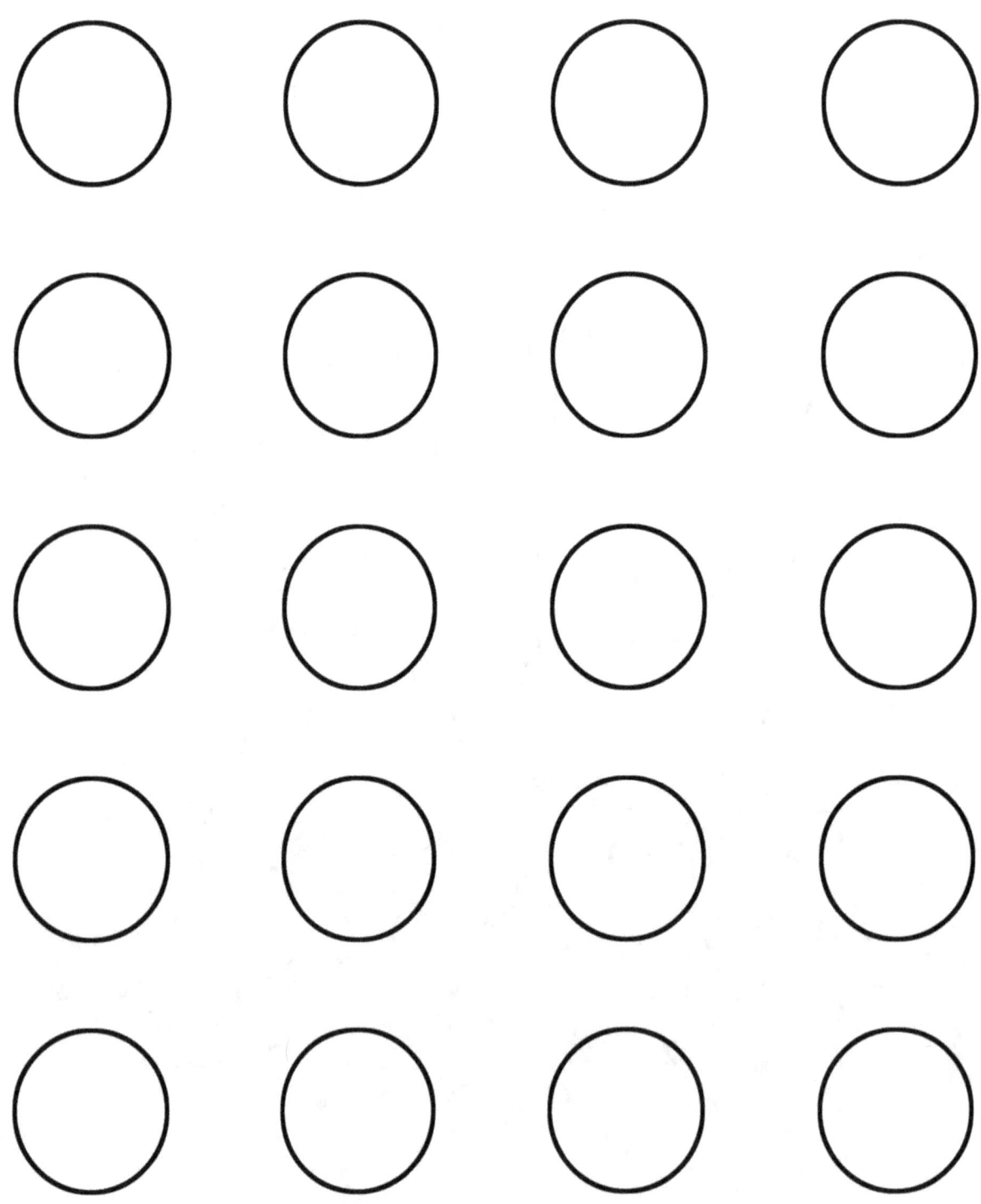

Test Your Color

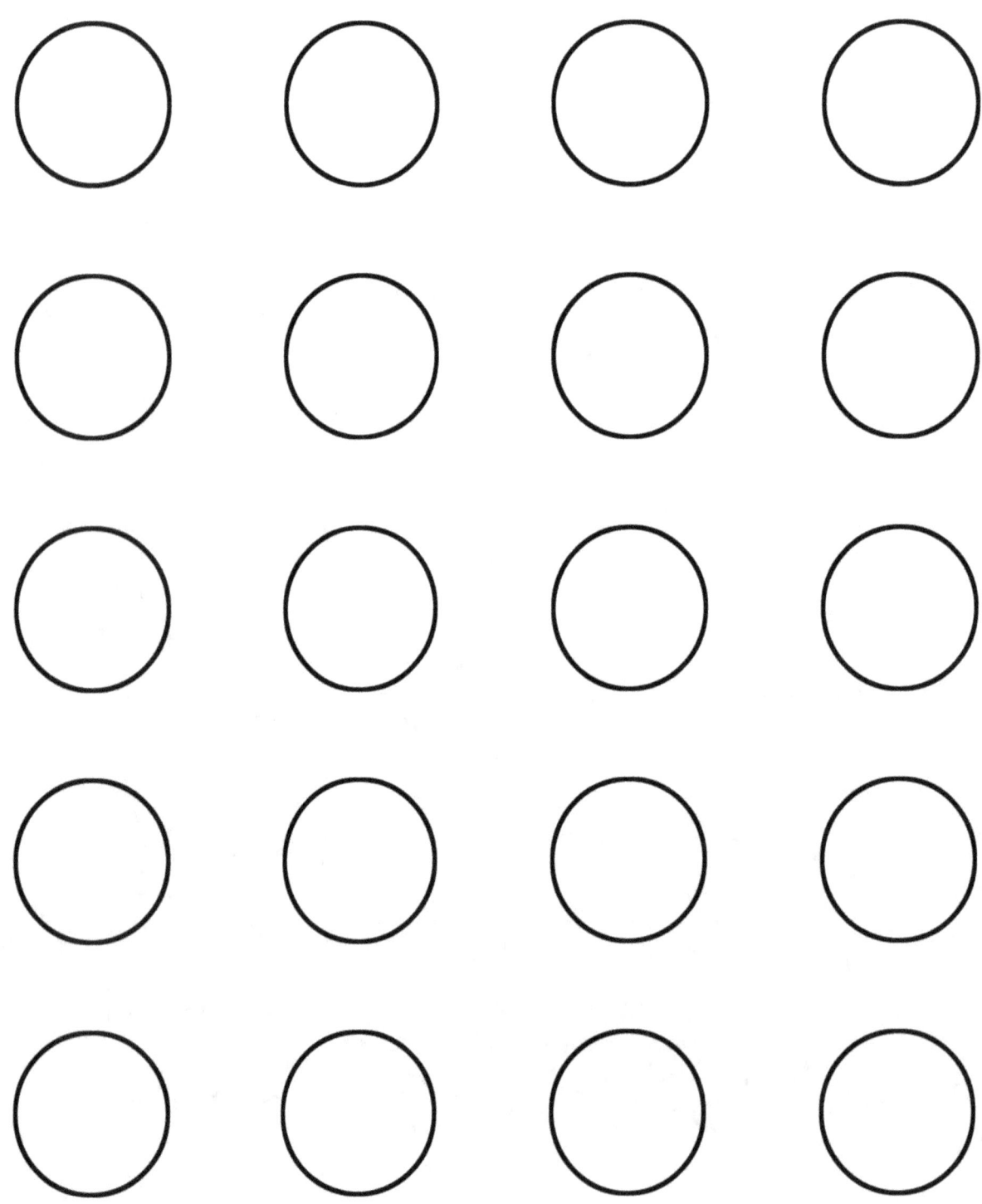

Draw a line from dot number 1 to dot number 2, then from dot number 2 to dot number 3, 3 to 4, and so on. Continue to join the dots until you have connected all the numbered dots. Then color the picture!

Test Your Color

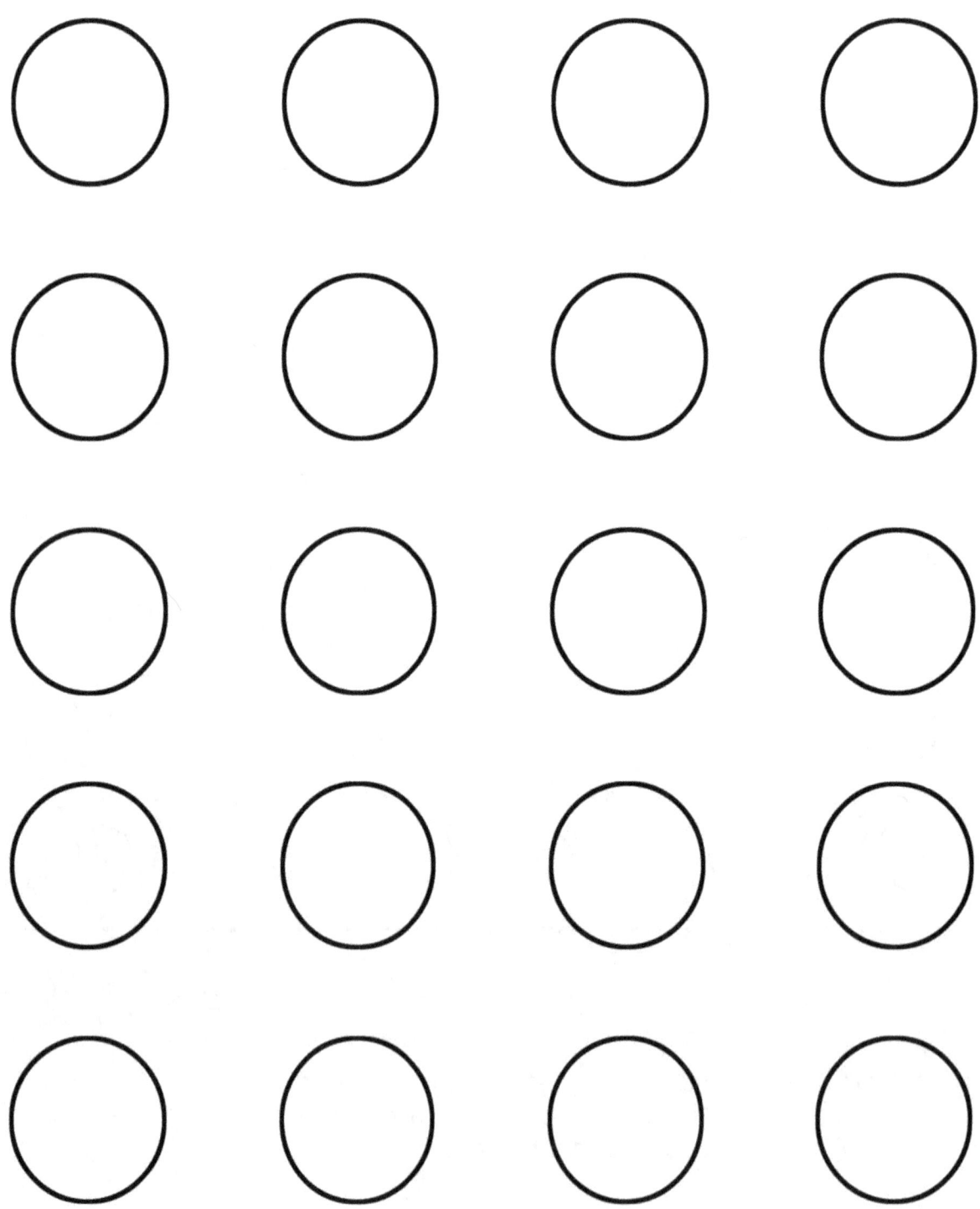

Draw a line from dot number 1 to dot number 2, then from dot number 2 to dot number 3, 3 to 4, and so on. Continue to join the dots until you have connected all the numbered dots. Then color the picture!

Test Your Color

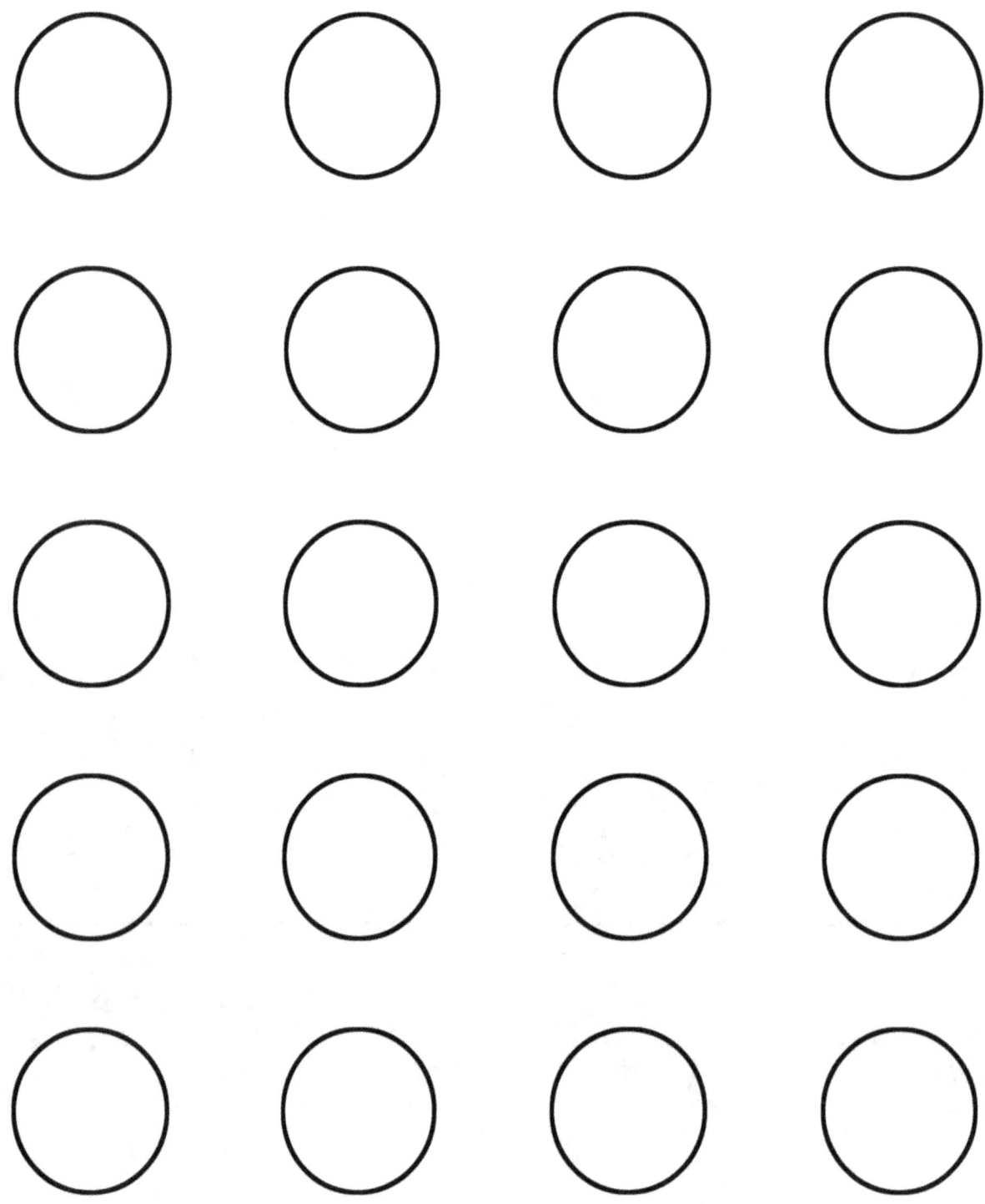

Draw a line from dot number 1 to dot number 2, then from dot number 2 to dot number 3, 3 to 4, and so on. Continue to join the dots until you have connected all the numbered dots. Then color the picture!

Test Your Color

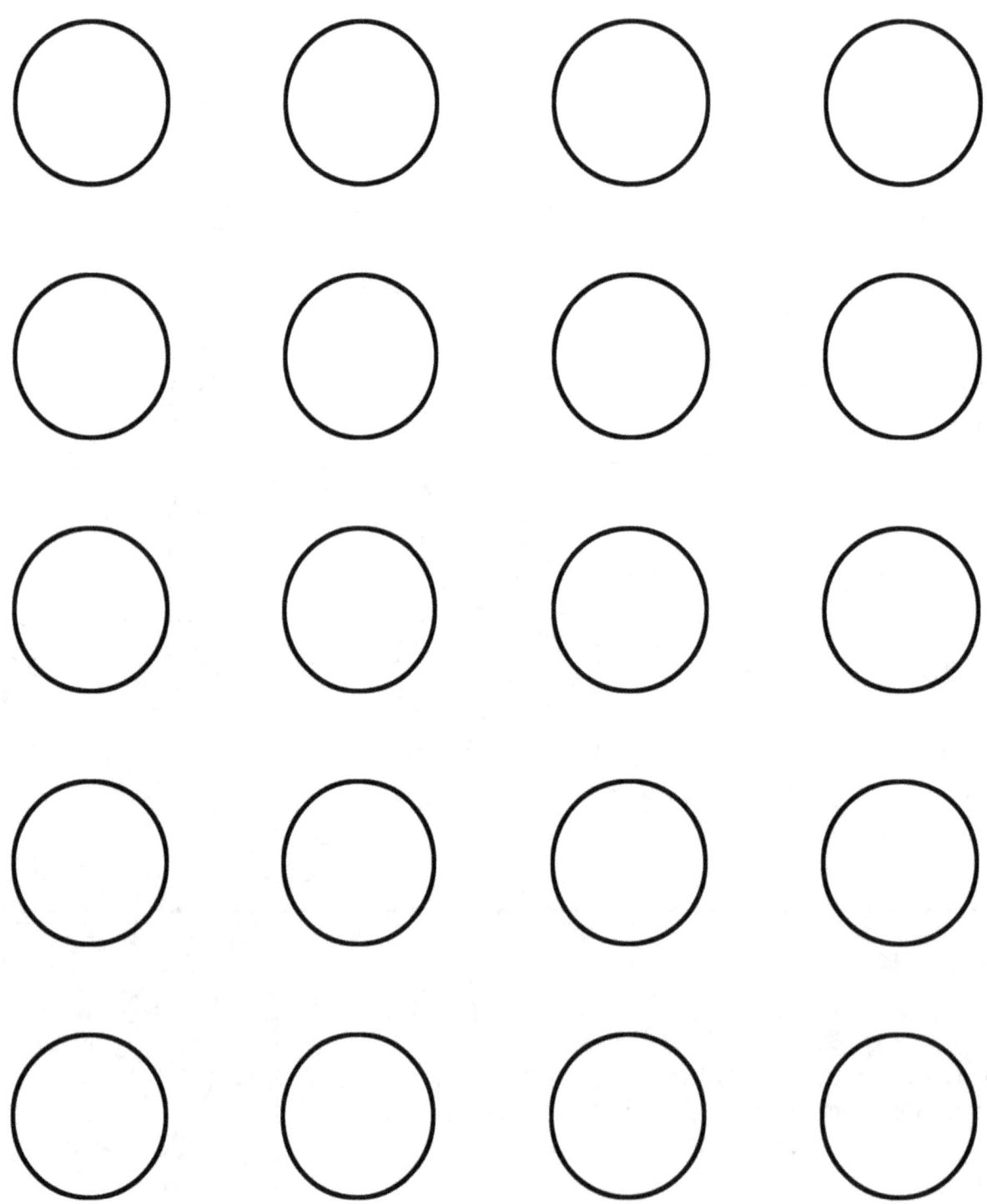

Draw a line from dot number 1 to dot number 2, then from dot number 2 to dot number 3, 3 to 4, and so on. Continue to join the dots until you have connected all the numbered dots. Then color the picture!

Test Your Color

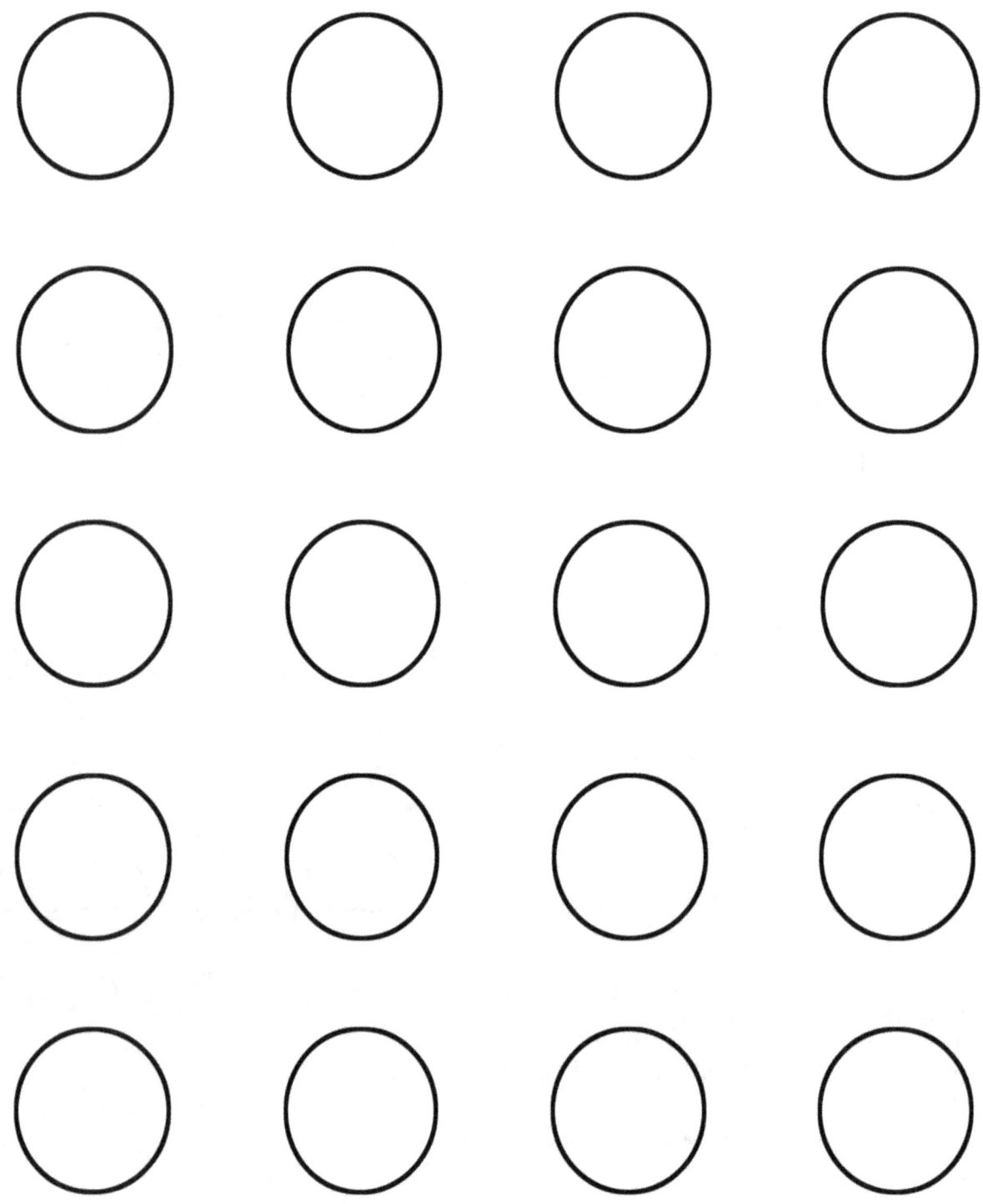

Draw a line from dot number 1 to dot number 2, then from dot number 2 to dot number 3, 3 to 4, and so on. Continue to join the dots until you have connected all the numbered dots. Then color the picture!

Test Your Color

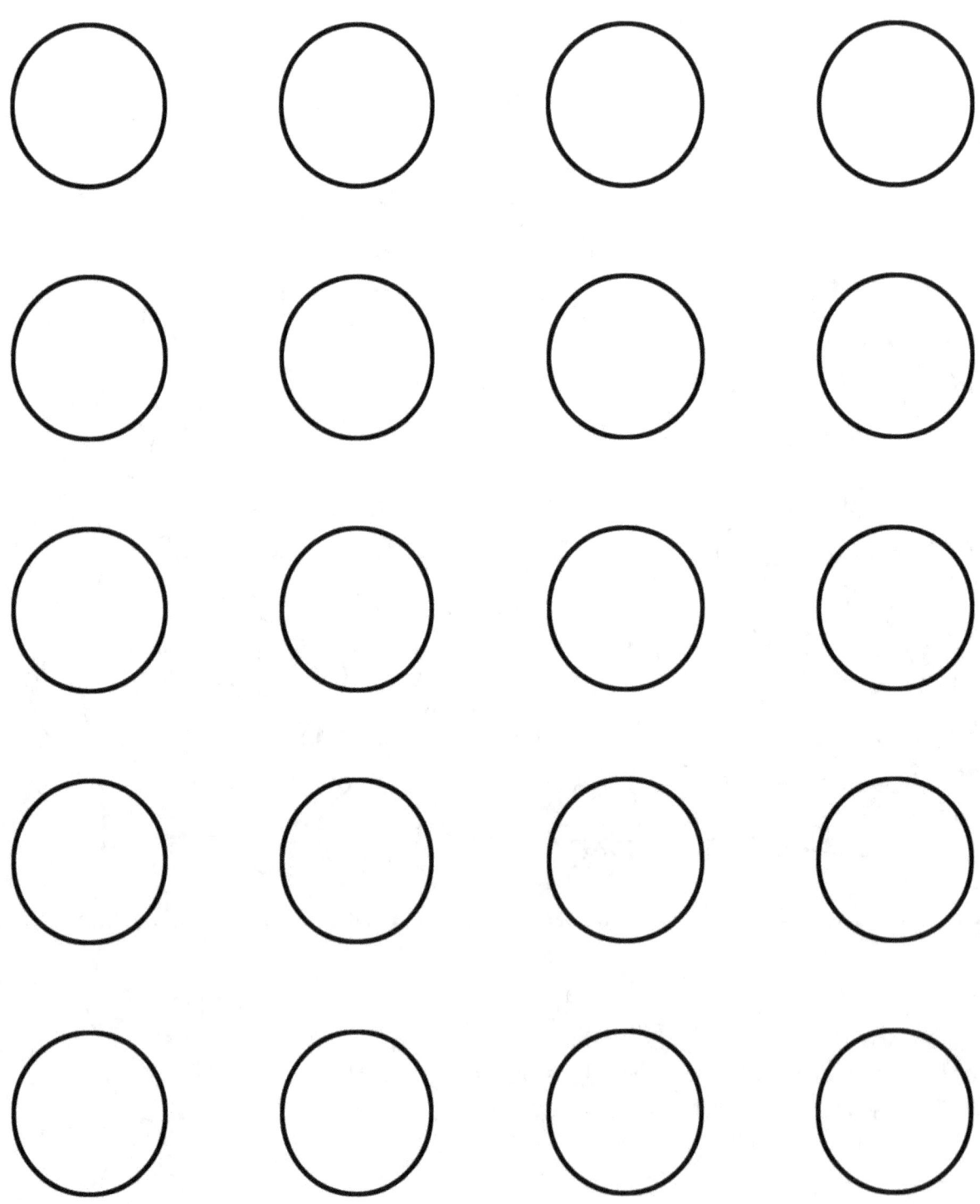

Draw a line from dot number 1 to dot number 2, then from dot number 2 to dot number 3, 3 to 4, and so on. Continue to join the dots until you have connected all the numbered dots. Then color the picture!

Test Your Color

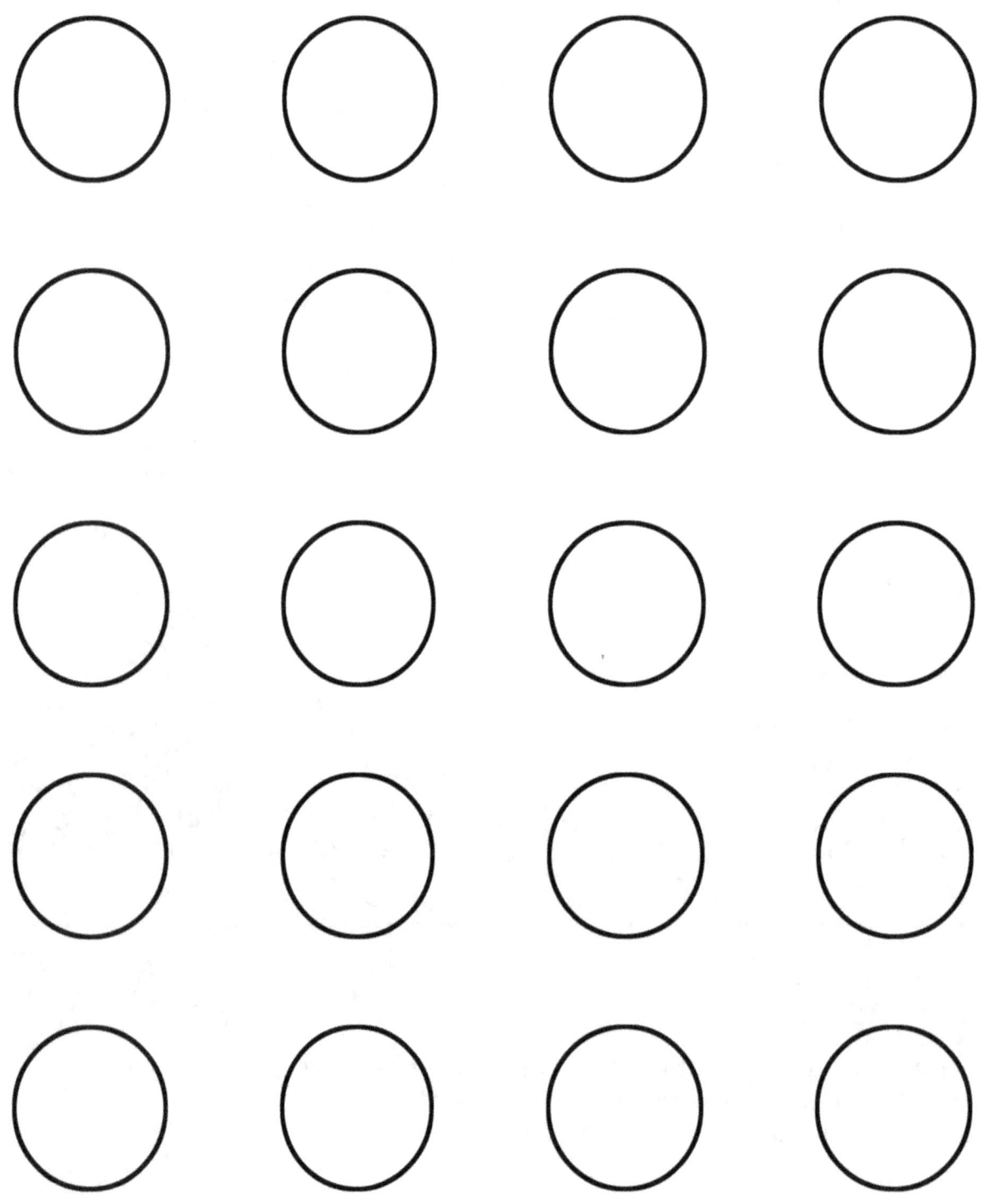

Test Your Color

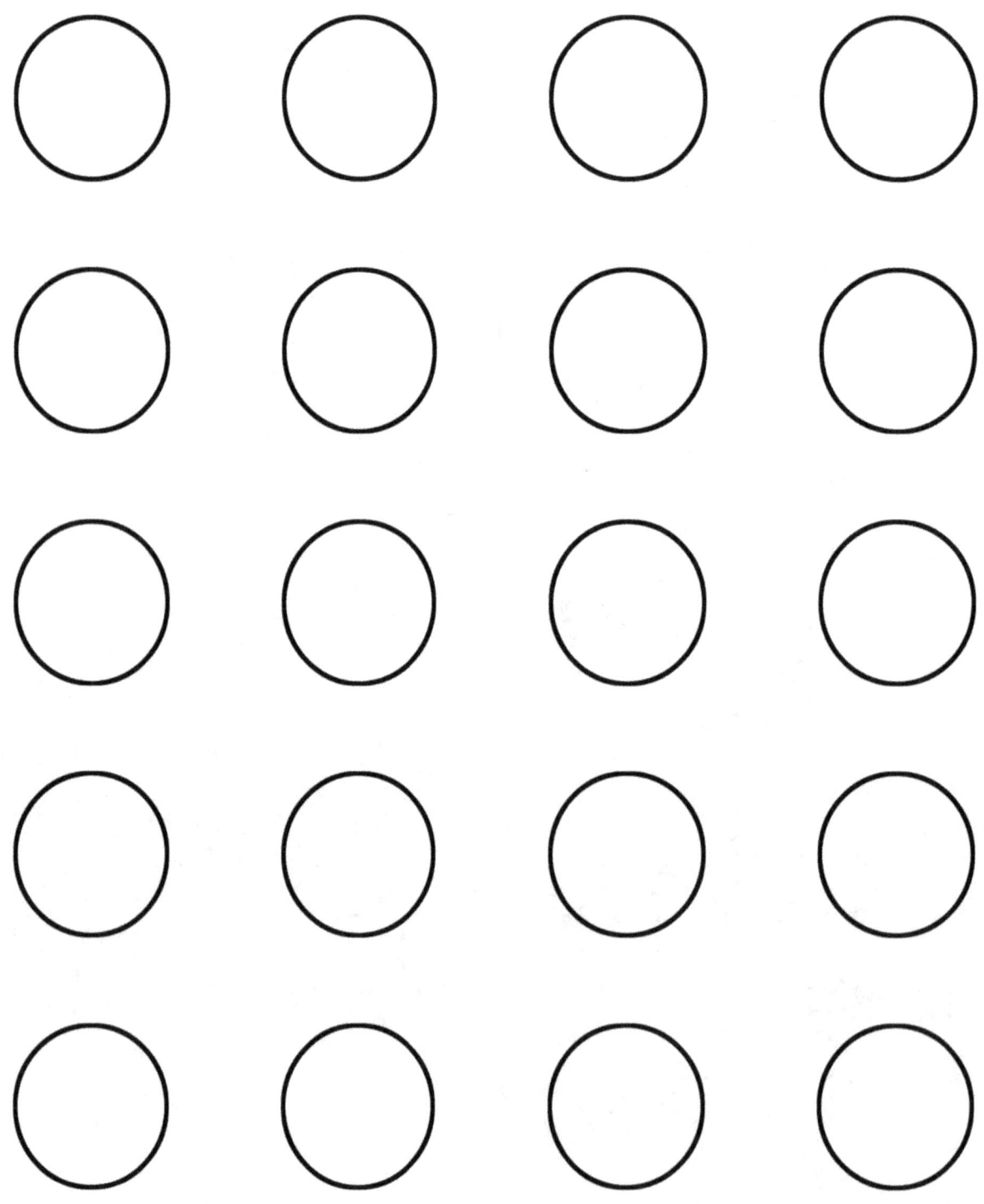

www.ingramcontent.com/pod-product-compliance
Lightning Source LLC
Chambersburg PA
CBHW081425220526
45466CB00008B/2280